T0391189

Ceramic Processing

Ceramic Processing

R.A. Terpstra

Manager of the Centre for Technical Ceramics
Netherlands Organization for Applied Scientific Research
TNO, Eindhoven, The Netherlands

P.P.A.C. Pex

Project Manager, Membrane Materials Development
Netherlands Energy Research Foundation
ECN, Petten, The Netherlands

and

A.H. de Vries

Product Manager, Traditional Ceramics
Netherlands Organization for Applied Scientific Research
TNO, Eindhoven, The Netherlands

SPRINGER-SCIENCE+BUSINESS MEDIA, B.V.

First edition 1995

© 1995 Springer-Science+Business Media Dordrecht
Originally published by Chapman & Hall in 1995
Softcover reprint of the hardcover 1st edition 1995

Typeset in 10/12 pt Times by Thomson Press (India) Ltd., New Delhi
Printed in Great Britain by St Edmundsbury Press, Bury St Edmunds

ISBN 978-94-010-4236-9   ISBN 978-94-011-0531-6 (eBook)
DOI 10.1007/978-94-011-0531-6

A catalogue record for this book is available from the British Library

Library of Congress Catalog Card Number: 95-67913

♾ Printed on acid-free text paper, manufactured in accordance with
ANSI/NISO Z39.48-1992 (Permanence of Paper).

Contents

Contributors

R.A. Bauer, PhD, Department of Inorganic Materials, TNO, Eindhoven, The Netherlands.

D. Bortzmeyer, PhD, Leader of Shape Forming Group, Rhône-Poulenc Research, Aubervilliers, France.

R.G. Horn PhD, Professor of Materials Science, University of South Australia, Adelaide Australia.

M.A. Janney, PhD, Metals and Ceramics Division, Oak Ridge National Laboratory, Tennessee, USA.

I.J. McColm, DSc, DPhil, FRSC, C. Chem., FIM, FRSA, Professor of Ceramic Materials, Department of Industrial Technology, University of Bradford, UK.

R.E. Mistler, ScD, Richard E. Mistler Inc., Morrisville, Pennsylvania, USA.

J. Schoonman, MSc PhD, Professor of Applied Inorganic Chemistry, Delft University of Technology, The Netherlands.

Foreword

Two years after the foundation of the European Ceramic Society the first ECerS Conference was held in Maastricht, The Netherlands. After this successful conference it was felt that the ECerS should not only stimulate ceramic research but also run projects of an educational character. To this end the Netherlands Ceramic Society decided to sponsor a summerschool on ceramic processing intended for young ceramists from European countries. Several internationally recognized ceramic specialists were prepared to lecture and the course was fully attended. In fact, this course showed that there is a clear need for dedicated courses at a moderate fee. A venue was chosen where the participants would have ample opportunities for personal contacts. In this way the summerschool gave the students more than purely technical training.

After the meeting the ECerS Council established an Educational Committee to take care of educational matters on a permanent basis. Following an initiative of the French Ceramic Society a special Conference of Higher Educational Institutions in Ceramics was organized under the name Euroforum in 1992 in Limoges, France. The second Euroforum was held in 1994 in Höhr-Grenzhausen, Germany. These meetings aim to bring together educators, graduates and industrial managers. Further activities under the auspices of the ECerS are planned.

Having been involved in the preparation of the course, it is a great pleasure to introduce the book of the proceedings of the NKV-summerschool. The contents and the level are such that we trust that many ceramists will benefit from the book.

I congratulate the editors and organizers of the summerschool on their initiative.

R. Metselaar
Past-president, ECerS

Preface

This book has been edited from the lectures given at the NKV-summerschool on ceramic processing which was organized by the Netherlands Ceramic Society (NKV), 5–9 September 1991 in Petten, The Netherlands.

The NKV-summerschool was sponsored by:

- The Netherlands Ceramic Society, NKV
- The European Ceramic Society, ECerS
- Ceramic Manufacturers Association, VKI
- The Netherlands Energy Research Foundation, ECN
- The Netherlands Organisation for Applied Scientific Research, TNO
- The National Ceramic Centre, NKA
- The Centre for Technical Ceramics, CTK
- N.V. Royal Sphinx, Maastricht
- Hoogovens Industrial Ceramics, HIC
- Gimex, Geldermalsen
- NIFA Instruments, Leeuwarden
- CAM Implants, Leiden
- Royal Tichelaar, Makkum
- Ceratec, Haaften

The following people were responsible for organizing the NKV-Summerschool:

- Dr ir R.A. Terpstra, CTK (Chairman)
- Ir P.P.A.C. Pex, ECN (Secretary)
- Mr A.H. de Vries, TNO-Ceramics (Treasurer)
- Prof. dr R. Metselaar, CTK/Eindhoven University of Technology (Scientific Advisor)
- Mrs C.A.L. Ruitenburg, ECN (Local organizer)
- Ing. P.J. van Tilborg, ECN (Local organizer)

The NKV-summerschool was arranged to provide (graduate) students, scientists and engineers an opportunity to attend an advanced course in ceramic processing taught by internationally well-known lecturers.

The lectures were presented by specialists on topics most relevant for ceramic processing, starting from powder synthesis and the behaviour of powder particles in dispersions in relation to colloidal consolidation processes, various forming routes like dry pressing, tape casting, injection moulding and extrusion. The concluding lecture, which could not be included in this book, treated the theoretical aspects of the sintering process. The opening lecture provided an overview of the trends in ceramic materials and their processing, answering various questions like: why are a limited range of new materials of current interest?, which may become dominant?, what are they?, why develop ultrafine powders?, and what are the prospects for ceramic fibres?

Since the lectures provided an excellent comprehensive overview by specialists in the field of ceramic processing, it was suggested that the organizers publish a book out of the proceedings of this summerschool. The result is this volume on the most important aspects of ceramic processing written by seven specialists in the field.

R.A. Terpstra
P.P.A.C. Pex
A.H. de Vries

Eindhoven/Petten

CHAPTER 1

Special ceramics for modern applications: which? why? how?

I.J. McColm

1.1 INTRODUCTION

This chapter presents an overview of the trends in ceramic materials and in their processing. Why are a limited range of new materials of current interest? Which may become more dominant? What are they? Why develop ultrafine powders? Do ceramic fibres have worthwhile properties compared to the effort needed to produce them? Can realistic cost targets be achieved? These questions will be approached by examining the concepts of specific strength, specific stiffness, theoretical strength, surface energy, fracture toughness, critical length, pull-out work, limiting particle size and green structure linked to examples of production methods.

In the 1960s ceramic manufacture was centred mainly in Europe and Japan, with Japan producing cheap reproductions of European wares. Now the Japanese ceramic industry is the largest in the world and many countries, such as Korea, Thailand and China, have taken over as cheap production centres. When production becomes uneconomic with respect to certain ware and certain designs in Europe it is becoming common practice to have such items manufactured where it is cheap, whilst still maintaining the brand name.

This situation is one of the factors driving ceramic producers, ceramic scientists, and ceramic technologists towards new technology and new materials. However, within the traditional industry this is represented by changes introduced to lower the costs of existing technology, rather than looking towards completely new products. This is a short-sighted policy: the ceramics industry needs to follow the lead of other technologies.

For example, the manufacture of plates by clay-dust pressing in a steel and nylon mould at high pressure was introduced to make plates faster and more accurately with less labour. This is seen as an advance on the use of soft clay in

Ceramic Processing. Edited by R.A. Terpstra, P.P.A.C. Pex and A.H. de Vries.
Published in 1995 by Chapman & Hall, London. ISBN 0 412 59830 2

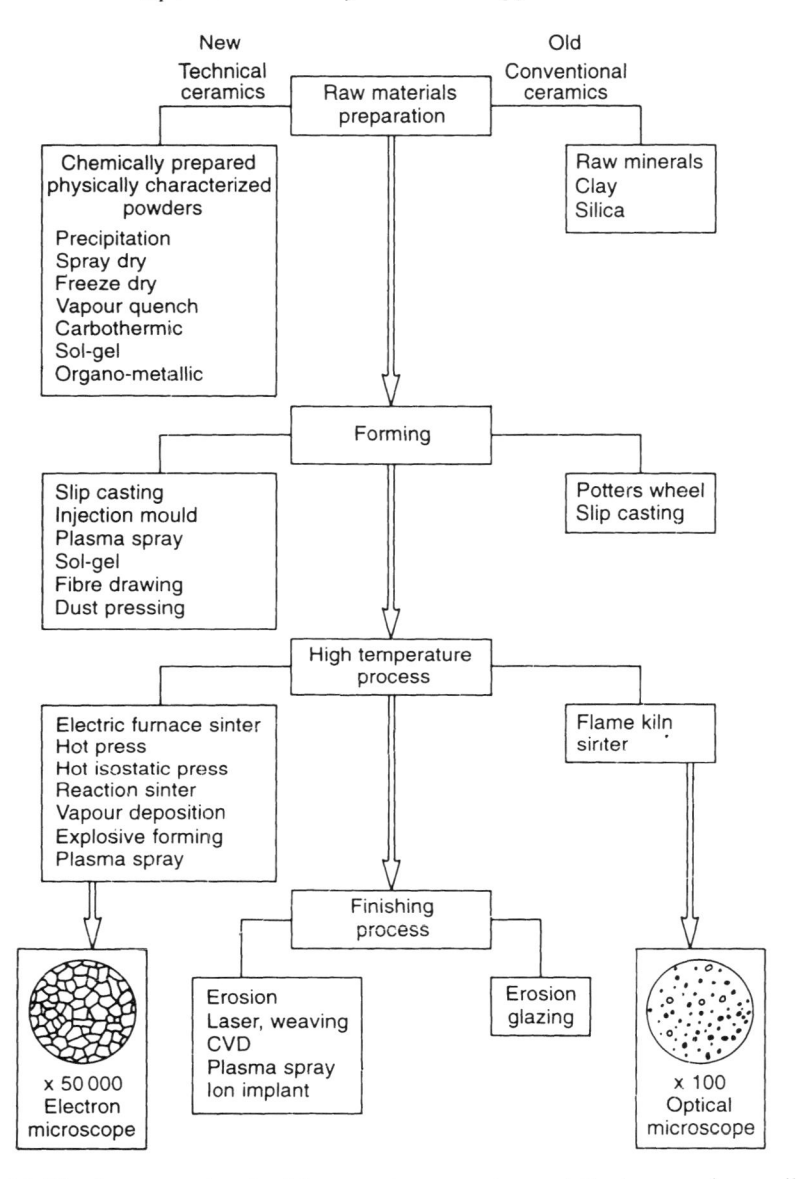

Fig. 1.1 The four steps involved in ceramic processing and the increase in possible ways of achieving them on moving from conventional to special ceramics.

a spinning plaster mould, which was itself seen as a development of the potter's wheel. It would be presented, however, as a use of techniques well established in the powder metallurgical industry, so perhaps a wider look at other industrial technologies would be rewarding for the ceramics industry.

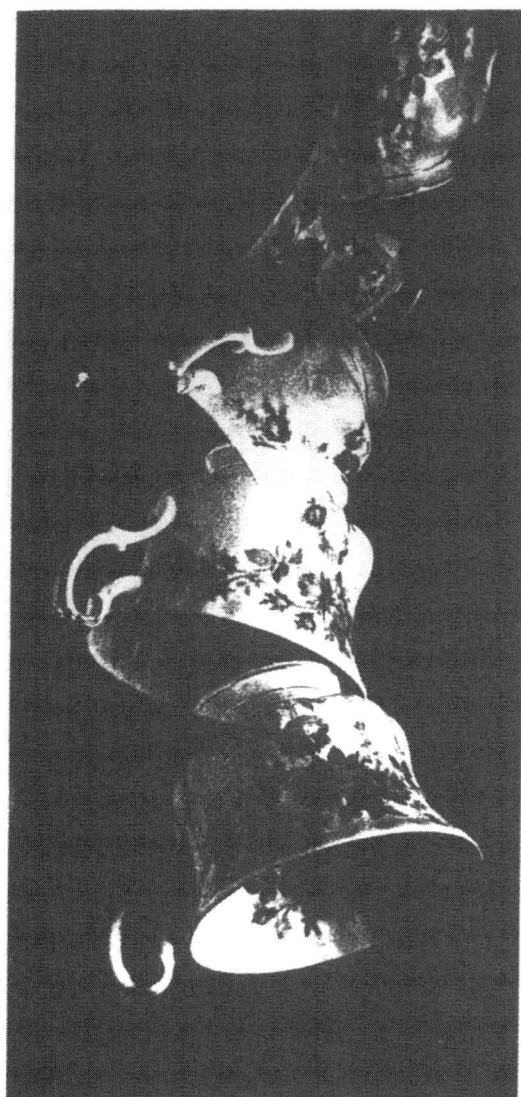

Fig. 1.2 A recent advertisement by the Alcoa company which emphasizes that traditional ceramics can be improved by researching into powder morphology and new forming processes for modern materials applications.

Evolution of Engineering materials

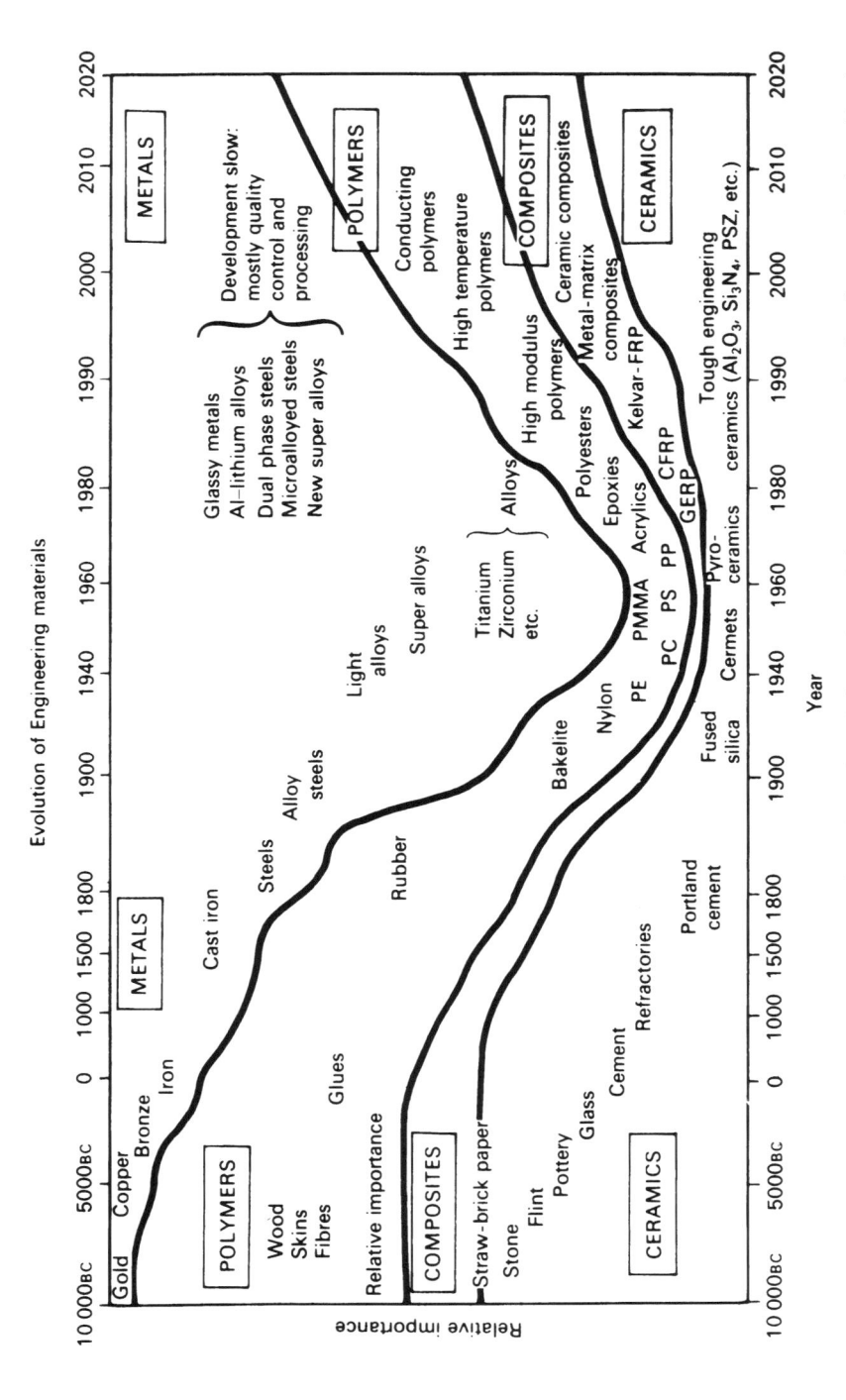

Fig. 1.3 Ashby's diagram [1] showing schematically how after a balanced material usage pattern technology became dominated by the use of metals. Beyond 1990 the figure predicts a return to more balanced usage with ceramics and ceramic-composites becoming more important again.

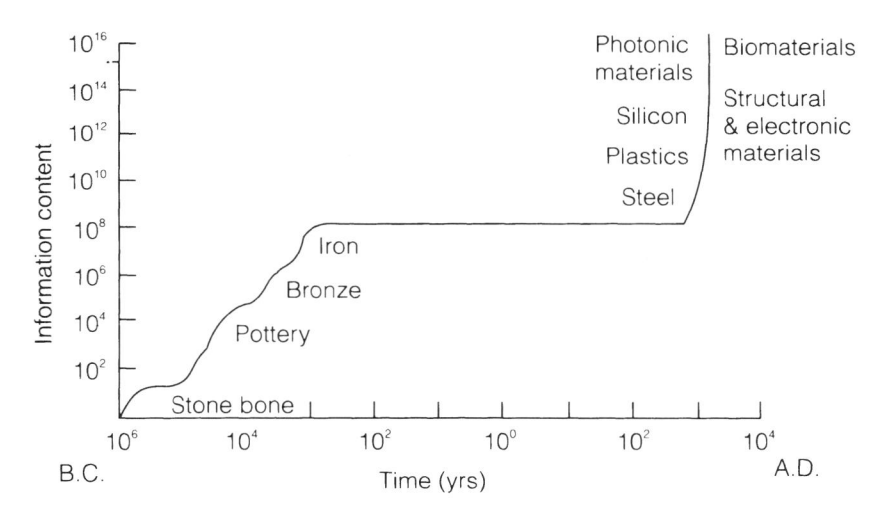

Fig. 1.4 A schematic diagram showing the growth in technology information content inherent in the use of new materials. It emphasizes the present rapidly changing circumstances before reaching another plateau when the second industrial revolution nears maturity.

The way to advanced high technology ceramics lies in the roots of several industries as shown in Fig 1.1. The unbreakable cup, as highlighted in a recent Alcoa advertisement (Fig. 1.2), was achieved by bringing together science and technology from several sectors.

The development and analysis of several simple equations and ideas can show us the way to successful developments in an industry which is now some 7000 years old. The field of ceramics and composites containing ceramics was once 50% or more of the materials experience of humanity and is set to rise to this again when metals diminish in importance, see Fig. 1.3. With such a change we are in the midst of exponentially rising increases in technology information content, following a long period of relative stagnation, and this is indicated in Fig. 1.4.

In a time of such rapid change can we make useful scientific and technological predictions? We will attempt to do this now using a few simple concepts and equations.

1.2 PREDICTIONS ARISING FROM SIMPLE CONSIDERATIONS OF STRENGTH

Clearly, to obtain the unbreakable cup shown in Fig. 1.2 a choice of material of ultimately exceptional strength should be made, and so here we look to the fundamental very early work of Griffith [2] and Orowan [3].

1.2.1   Group 1: flawless material

The method outlined below is basically that developed by Orowan; no realistic model of structure is involved. When a tensile stress is applied, the force is opposed by bonding forces between sheets of atoms perpendicular to the applied stress. In the region of small extensions, that is the elastic region, the applied force and the restoring forces in the chemical bonds are equal and opposite.

If the original intersheet distance is a_o and the extension in the direction of applied force is x, then

$$\sigma_{app} = Ex/a_o. \tag{1.1}$$

In equation (1.1) E is Young's modulus and x and a_o are defined above. A problem now arises in that equation (1.1) only applies to very small extensions. If the model is to extend to total intersheet separation then we cannot expect E to remain constant. For example, ionic bond forces are proportional to $1/r^2$ and covalent bond forces are exponentially related to separation.

Nevertheless, it is still worth pursuing Orowan's logic but bearing these points in mind.

If the applied stress is now increased, the restoring stress due to the bond forces also increases and this situation continues until the maximum value, σ_m, is reached; this we can call molecular strength or theoretical strength, and once $\sigma_{app} > \sigma_m$ breakage occurs.

A form for the restoring stress has to be postulated, and from the many possibilities a simple sine function can be chosen

$$(\sigma_{restoring}) = \sigma_{app} = \sigma_m \sin 2\pi x/\lambda. \tag{1.2}$$

In equation (1.2) λ is a distance less than an interplanar separation.

We see from equation (1.2) that when the planes are extended by $x = \lambda/4$, the restoring force will be at its greatest, i.e. σ_m, and when x is extended to $\lambda/2$, the prediction is zero attractive force, and planar separation, that is, fracture, occurs.

For small values of x, $\sin x = x$ and equation (1.2) differentiates to:

$$d\sigma_{app}/dx = 2\pi\sigma_m/\lambda. \tag{1.3}$$

Equation (1.1) differentiates to

$$d\sigma_{app}/dx = E/a_o. \tag{1.4}$$

From equation (1.3) and equation (1.4), $2\pi\sigma_m a_o = E\lambda$.

Hence, the theoretical strength is given by equation (1.5)

$$\sigma_m = E\lambda/2\pi a_o. \tag{1.5}$$

This is almost useful, except that it contains the unknown λ, and so the next stage is to eliminate it by using the fact that on fracture two new surfaces are formed, each with surface energy per unit area, γ, that must be supplied.

Hence, the work done per unit area by the applied stress must equal 2γ, assuming that all other energy losses such as sound, heat, etc. are negligible. From equation (1.2), fracture occurs at $x = \lambda/2$ hence

$$\int_0^{\lambda/2} \sigma \, dx = \int_0^{\lambda/2} \sigma_m \sin 2\pi x/\lambda \, dx = \lambda \sigma_m/\pi = 2\gamma. \tag{1.6}$$

Equations (1.5) and (1.6) can be combined to eliminate λ thus arriving at an expression for the theoretical strength

$$\sigma_m = (E\gamma/a_o)^{\frac{1}{2}}. \tag{1.7}$$

This expression is quite general in that no structural model or bonding arrangement was assumed, but just a mode of sudden catastrophic failure across a plane. It is useful for setting the scene, so to speak, and forms the basis for material comparisons.

Equation (1.7) suggests that we need materials with:

- **High Young's modulus.** This can only be achieved by utilizing materials with covalent bonding and, as a second choice, materials with ionic bonding since in general

$$E_{cov} > E_{ionic} > E_{metal} \gg E_{van\ der\ Waals}.$$

- **Small interplanar separation a_o.** Another way of saying this is close packed structures. Here we have the conflict that covalent structures are open networks in general and so have relatively large a_o values compared to ionic and metallic solids.
- **A large surface energy, γ.** It is difficult to measure this property or to assess its relative value.

So far we have only made limited progress in our quest to simplify prediction. More progress can be made by replacing the surface energy term, γ, by more accessible terms.

1.2.2   Replacement of the surface energy term in equation (1.7)

Equation (1.7) contains the term γ, the excess surface free energy, which has to be large to give high intrinsic strength. However, it is not a frequently encountered property outside mechanics courses, and it is not immediately obvious which ceramics will have large values of γ. Therefore it is instructive to derive an expression to replace γ in equation (1.7).

Consider a crystal containing N atoms of which N_s are at the surface, leaving $N - N_s$ as internal atoms. We now make the grossly simplistic assumption that only near-neighbour atoms bond to each other, and the total binding energy, TBE, of the crystal can then be found by summing each bond. Every inside atom has twelve near neighbours, six in its own layer, three above and three below. Each surface atom has fewer near neighbours, eight if the unit cell is face centred cubic (f.c.c.). Hence

$$\text{TBE} = \tfrac{1}{2}\Delta E[12(N - N_s) + 8N_s] = \tfrac{1}{2}\Delta E[12N - 4N_s]. \tag{1.8}$$

In equation (1.8) ΔE is the binding energy between any two near-neighbour atoms, which explains the factor $1/2$, without it equation (1.8) would be counting each interaction twice. Now a thought experiment can be done: what would the TBE be if the solid had no surface? The answer is expressed in equation (1.9)

$$\text{TBE} = \tfrac{1}{2}\Delta E 12N. \tag{1.9}$$

Clearly the difference between equations (1.8) and (1.9) is the effect a surface has on the solid. Thus the solid has a surface energy of $-2\Delta E N_s$, and since the bond energy ΔE is negative, this is a positive term, meaning that the presence of a surface raises the energy of the system, making it less stable. Thus production of new surface absorbs energy, a fact used in fracture mechanics. The units of surface tension and specific free energy are kJ m^{-2}, 'specific' referring to unit area. For liquids the two quantities are numerically equal, but not for solids; atoms in a solid surface are not usually in equilibrium positions after cleavage because of their low mobility. This is partial progress, for we can now write

$$\gamma A = 2\Delta E N_s. \tag{1.10}$$

For $1\,\text{cm}^2$ containing z, the number of atoms per square centimetre, we can write equation (1.10) as

$$\gamma = 2\Delta E z. \tag{1.11}$$

Two steps are now necessary, first to replace z, and then ΔE, by using a model crystal. To replace z we must consider the volume of a f.c.c. solid; the side of the unit cell $2a$ produces a cell volume $8a^3$, and since the cell contains four atoms, the volume per atom is $2a^3$. If the material has a molecular weight M, then one atom or molecule has mass M/N_o, where N_o is the Avogadro constant. If the density of the material is ρ then $M/\rho N_o = 2a^3$ from which

$$a^3 = M/2N_o\rho. \tag{1.12}$$

The face of the unit cell, which now represents a surface, has an area $4a^2$ and contains two atoms. Therefore each atom occupies an area $2a^2$.

$$z = \text{number of atoms per cm}^2 = \frac{1}{2a^2}$$

which from equation (1.12) leads to

$$z = \tfrac{1}{2}(2N_o\rho/M)^{\frac{2}{3}}. \tag{1.13}$$

In equation (1.13) we have expressed z in terms of the readily comprehended and assessed quantities of mass and density.

To express ΔE in more meaningful terms we return to the model. What must be supplied if one internal atom is removed to the atmosphere? The answer is clearly six units of ΔE, since it has twelve nearest neighbours. When one mole of a solid is sublimed, the energy needed is ΔH_{sub}, and so one atom would require $\Delta H_{sub}/N_o$. Therefore $\Delta H_{sub}/N_o = 6\Delta E$ and

$$\Delta E = \Delta H_{sub}/6N_o. \tag{1.14}$$

Putting equations (1.14) and (1.13) into equation (1.11) gives

$$\gamma = 0.27\Delta H_{sub}\rho^{\frac{2}{3}}/N_o^{\frac{1}{3}}M^{\frac{2}{3}}. \tag{1.15}$$

Substitution of equations (1.15) and (1.12) into equation (1.7) gives the required equation (1.16)

$$\sigma_m = (0.33\Delta H_{sub}\rho E/M)^{\frac{1}{2}}. \tag{1.16}$$

Equation (1.16) is interesting because it leads to the criteria that a material must fulfil in order to have high intrinsic strength. Perhaps the most important and unexpected prediction is that suitable material must have a low molecular weight. This means that we can restrict further consideration to the first two rows of the periodic table of the elements.

1.3 MATERIALS PREDICTIONS BASED ON EQUATION (1.16)

From the work so far then we can look for covalent and ionic ceramic materials in the elements, binaries and ternaries found in the series:

H	He	Li		Be	B	C	N	O	F
	Ne	Na		Mg	Al	Si	P	S	Cl	Ar.

We can use the criteria established so far:

- low molecular weight;
- covalent, ionic or ionic + covalent bonding;
- high melting and boiling points to achieve high ΔH_{sub};
- high density.

Applying these criteria establishes some 20–30 materials of importance in an order of priority: C, B, Be, Si, B_4C, SiC, BeO, B_2O_3, Si_3N_4, SiAlONs, AlN, Al_2O_3, AlB_2, BN, $(NC)_x$, etc. Most of these, but not all, are at present being researched, developed and used to a differing degree.

There is, however, another set of criteria to be applied to the selection of ceramics for future development, namely:

- the product triangle, and
- achievement of forms of these materials which can be fabricated into the ideal microstructures able to give strength values approaching the predicted theoretical values.

1.3.1 The product triangle

The general properties of ceramics, and in particular, the specific properties of strength/density, modulus/density etc., of what are called the engineering or special ceramics, are now attracting considerable attention. Such specific properties are found in materials of widespread availability and potentially low cost so that it seems difficult to understand any reluctance to make and exploit them.

However, these attractive properties are only two-thirds of what can be described as the product triangle depicted in Fig. 1.5. That is, such properties are not the complete key to the adoption of new materials; the third apex of the triangle is fabrication technique, and without this there is no key to unlock the treasure chest of a high-technology future. The infrastructure of fabrication must be established if a new material is to become a commercial success. New materials usually involve development of fabrication routes, but these are very rarely completely new concepts, rather developments of techniques used to fabricate more established materials; modern ceramics are no exception to this pattern.

For centuries, ceramic fabrication technology and its related science has remained relatively simple, as the four stages of the process schematically shown in Fig. 1.1 were undertaken within the few headings shown on the right-hand side of the figure. As shown, these traditional processing steps lead to microstructures found to be acceptable only for traditional ceramics products. Rapid introduction of more complex operations to the four-stage process has been necessary, with the result that the choice of overall process can be quite bewildering, as the topics grouped on the left-hand side of Fig. 1.1 show. Much of the development in fabrication technology and science, therefore, has been

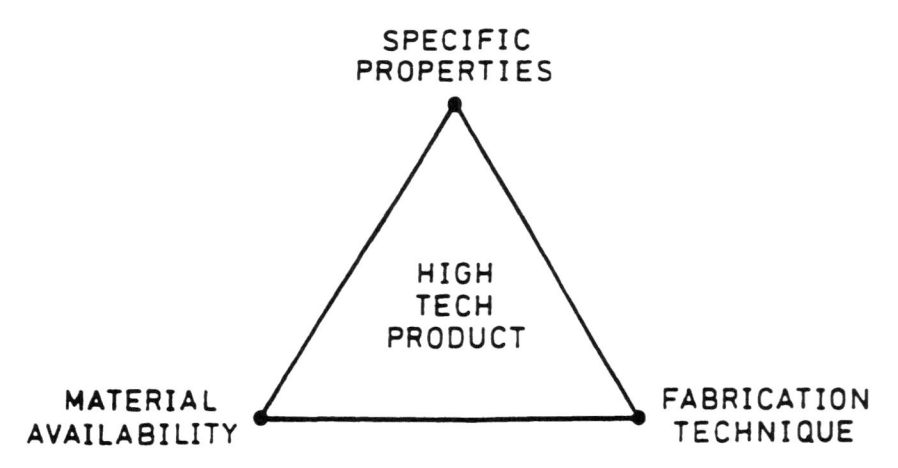

Fig. 1.5 The product triangle which demonstrates the important role of fabrication technique in the development of high technology products [4].

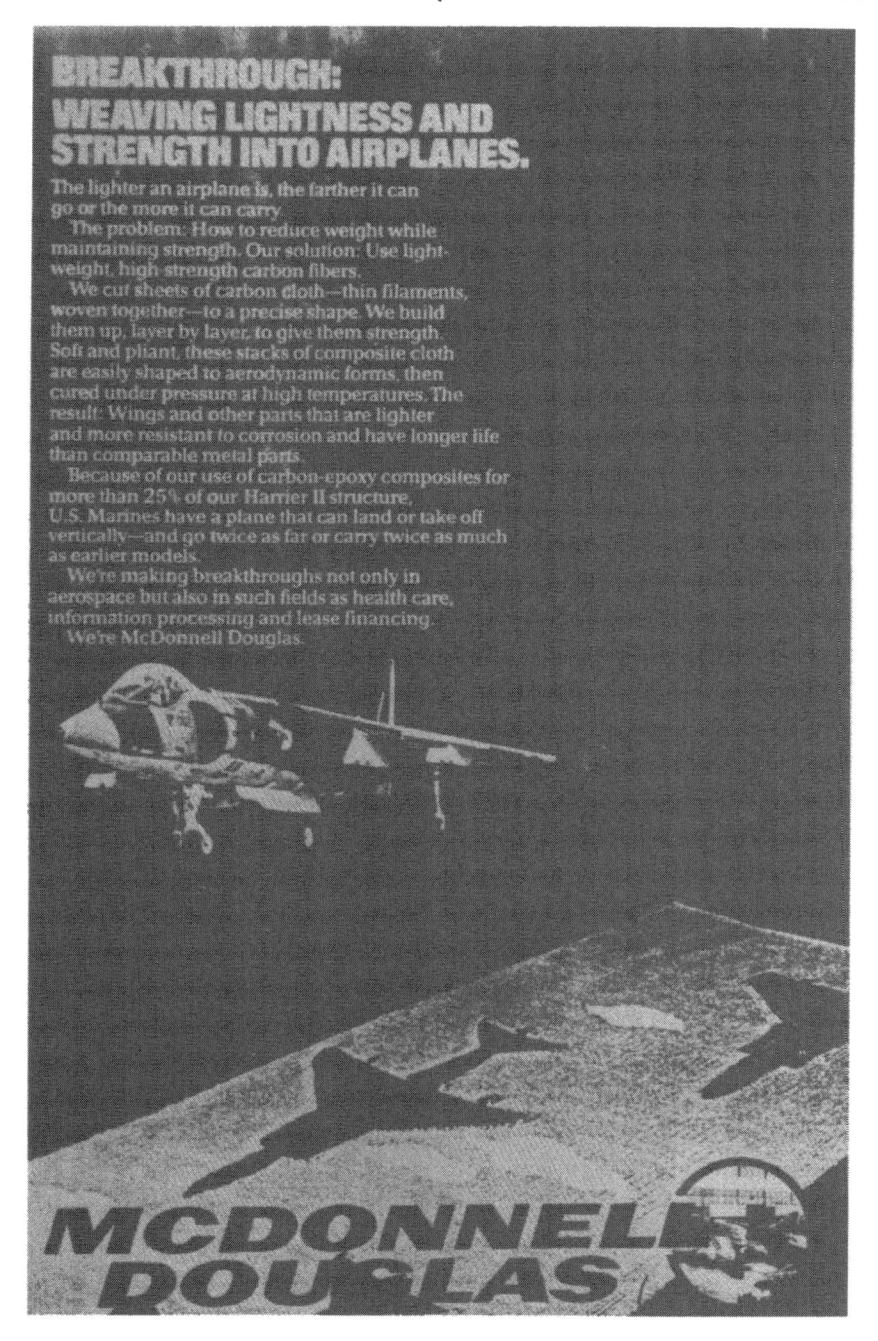

Fig. 1.6 An advertisement by the McDonnell Douglas corporation which emphasizes the development of new manufacturing routes in modern industries using ceramic components [5].

borrowed and developed from other areas of technology and science; ceramics has been learning from metallurgy, polymer science, physics and, not least, chemistry. In the future it is quite likely that this process will be continued by learning from textile technology, as ceramic based composites become important. The interdisciplinary development of these areas is valuable and necessary because the narrow approach of one group of contributors can lead to abnormal emphases and slow progress. For example, the development of sintering theory has relied heavily on physics. The result of this has been models that predict the rate of sintering or densification for specific materials at particular conditions with little broad-based predictability, yet this subject is the core of ceramic processing.

One purpose of this book is to develop an interdisciplinary approach and to group the principles from which the fabrication routes are emerging as techniques are borrowed and extended from other industries and disciplines.

Figure 1.6 demonstrates that at least in the USA and in the high technology defence industry a substantial move to relevant fabrication techniques involving the ceramic materials carbon, boron carbide and silicon carbide has occurred. Weaving and spinning will not be the only new fabrication techniques in the ceramic, and ceramic based, industries as we might see next by considering a simple equation for fracture toughness.

1.4 SOME PREDICTIONS BASED ON FRACTURE TOUGHNESS

The simple statement of the fundamental relationship of fracture mechanics is equation (1.17)

$$\sigma_B = K_{IC} a_c^{-\frac{1}{2}}. \tag{1.17}$$

This equation is a hypothesis predicting a linear relationship between the fracture stress, σ_B, and the reciprocal of the root of the length, a_c, of the fracture inducing defect. It has the importance of being capable of demonstrating how the strength of a ceramic component can be enhanced in ways that lead to developments in ceramic fabrication technology.

First, it tells us to improve the **damage resistance** term K_{IC}. This can be done by:

- Homofibre reinforcement, in which high quality material is made into fibres and used to strengthen a matrix of less well-formed material of the same composition. This causes cracks to branch around fibres as well as requiring energy, known as pull-out work, to be supplied.
- Ductile inclusions: the cermets.
- Stress-induced localized martensitic type transitions to phases of greater molar volume; zirconia and some carbides are typically used.

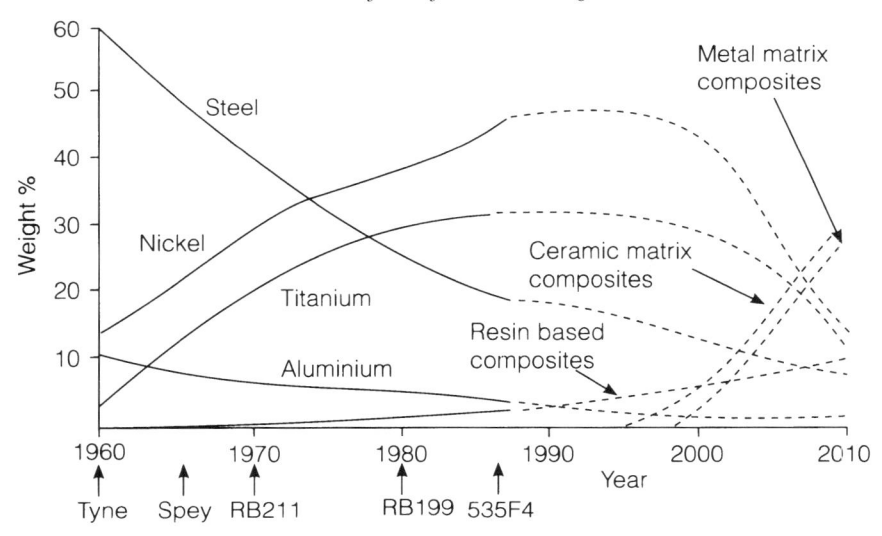

Fig. 1.7 A graph, supplied by courtesy of Rolls–Royce Ltd., demonstrating the trends in jet engine material usage which demonstrates the growth of ceramic usage via composites.

- Microcrack formation from anisotropic thermal expansion or incorporation of phases transforming locally from high to low density.

Figure 1.7 shows that in one high technology sector, exemplified by the jet engine makers Rolls–Royce, the first two methods above are expected to become more common towards the end of this century.

Second, from equation (1.17), we are directed towards reducing the size of defects through the $a_c^{-\frac{1}{2}}$ term. This implies:

- Producing uniform grain structures of ultra fine size. This is suggested in Fig. 1.1.
- Developing processing routes that lead to fewer process-related defects. In particular this means at the green state of the process.
- Proof testing to reject artefacts with $a_c > a_{max}$.

Third, as far as equation (1.17) goes, it is possible to minimize the effect of a_c and also improve K_{IC} by applying a finishing operation to the product. This means in effect to introduce a prestressed layer:

- Fibre reinforcement with tensile prestressing of the fibres from outside, or by choosing α_f not equal to α_{matrix} and E_f not equal to E_{matrix} where α is the thermal expansion coefficient and E is Young's modulus.
- Producing a compressively stressed surface layer either by glazing, by 'ion-stuffing', where small surface-layer ions are substituted by larger sized ones, or by utilizing a phase transformation to a lower density polymorph.

1.5 THE SIZE AND SHAPE OF MATERIALS NEEDED FOR PROCESSING

In section 1.2 we arrived at a theoretical strength equation that enabled us to see which are the potential super-ceramics. In section 1.4 we implied that flaw size limitation was important. Now in order to make a few more quantitative predictions we should start from an equation more generally observed for several classes of material but possessing the general form of equation (1.16).

$$(\sigma_m)^2 = EW/\pi(1 - v^2)a_c. \tag{1.18}$$

where v is Poisson's ratio.

The W term in equation (1.18) is a more general work-of-fracture term and replaces the surface energy.

So far we have suggested that if we make a piece of ceramic small enough we can suppress weakening due to critical defect initiated fracture. Now consider crack initiated fracture further by using equation (1.18).

Replace σ^2 by $(E/15)^2$ which is a reasonable upper value and taking v for a ceramic to be 0.25

$$E^2/225 = EW/\pi(1 - 0.0625)a_c$$

hence

$$a_c = 225W/0.93\pi E. \tag{1.19}$$

Typical values for E and W for Al_2O_3, for example, are $400\,\mathrm{GN\,m^{-2}}$ and $0.02\,\mathrm{kJ\,m^{-2}}$, which leads to an a_c value of about $0.004\,\mu\mathrm{m}$ for Al_2O_3 to be able to show its theoretical strength. Thus no part of the material must have a cross section as large as 4 nm or the critical crack length could occur.

Weakness by dislocation generation is another possibility. Ductile type fracture does not result directly from the movement of dislocations originally present, but rather a mechanism for dislocation generation is needed. A dislocation held at its ends is a potential source of other dislocations. This is the Frank–Read source.

For such a source the stress needed to generate dislocations is given by

$$\tau = 2Gb/L. \tag{1.20}$$

In equation (1.20) G is the shear modulus, b is the Burgers vector and L is the length of the dislocation.

Now assuming an orientation of the plane containing the dislocation, say $45°$ to the direction of applied tensile stress, to permit slip at the lowest possible stress

$$\sigma_{th} = 2\tau = 4Gb/L. \tag{1.21}$$

Again if $E/15 = \sigma_{th}$, and G is approximately $E/3$, since L is about $24b$ and since b is typically $0.25\,\mathrm{nm}$, then a fibre or grain of about 6 nm diameter would reach theoretical strength from this mechanism.

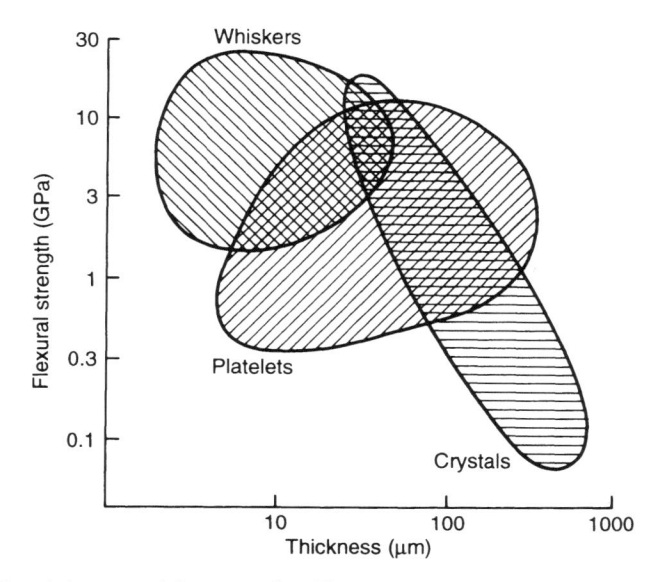

Fig. 1.8 Trends in strength increases for silicon carbide as a function of particle size and shape.

Although these dimensions seem impractical they can be approached in the form of whiskers and platelets which do, as Fig. 1.8 shows, have sharply increasing strength as their size diminishes.

The material form with greatest potential for high strength, coupled with high modulus, and low density, is the 'single crystal whisker'. The diameter of these is approximately 1 nm, and up to a few millimetres in length. Crystallographic planes make up their faces, so they have rectangular or diamond-shaped cross sections. They contain generally only one 'screw dislocation', near to their centre. Their surfaces are generally very smooth, which supresses fracture from surface flaws, but is not particularly good at inducing pull-out work and hence increasing toughness. Their strength can approach theoretical strength.

Another form of strong material is the 'platelet' or microplate. Several compounds can be made to grow in laminar form, mica is a well known example. Although potentially strong, as dislocations are not mobile within them, they mostly have smooth edges to achieve high strength.

The importance of size and shape is shown in Fig. 1.8 which shows the strength of various forms of SiC.

By the same arguments the particle sizes of powders which are to be sintered to a given shape must be less than the sizes calculated here and must not grow by grain growth mechanisms during the sintering process if the material is to show strength approaching its theoretical value. This is the driving force leading to the development of new powder manufacturing methods, see, for example, Chapter 2.

Why develop new powder production methods? One answer is that mechanical ways of making powders will not approach the small sizes needed as the next section shows.

1.6 PREDICTION OF LIMITING PARTICLE SIZE ARISING FROM MECHANICAL GRINDING

From an analysis of compression effects on cracked specimens it is possible to show that comminution by pressure becomes impossible for small particles once a critical size is reached. The model is given by Kendall and is developed from Fig. 1.9.

The axial cracking force, F in Fig. 1.9, is related to the material properties and the particle geometry, and so what follows related to the specific shape in Fig. 9.

$$F_{\text{crack}} = [B/(1-w/d)][2E\gamma_\pi d/3]^{\frac{1}{2}}, \qquad (1.22)$$

where the dimension terms are as defined in Fig. 1.9 with E the Young's modulus and γ_π the specific surface work. The $E\gamma_\pi$ term is the material-related parameter.

There will be an applied force capable of causing yielding, given by equation (1.23)

$$F_{\text{yield}} = \sigma_y Bw \qquad (1.23)$$

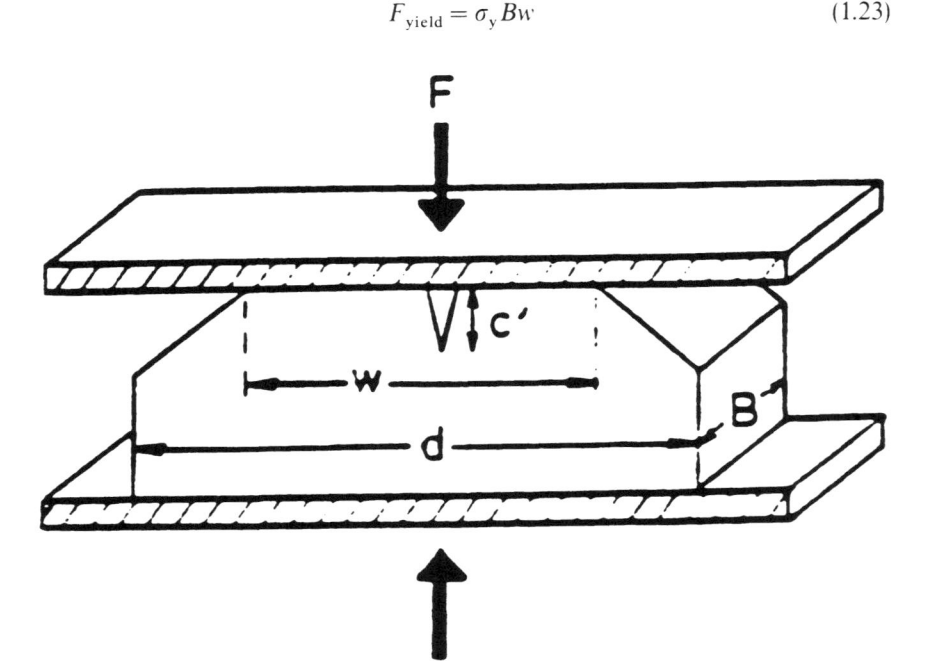

Fig. 1.9 The model used to develop the equations in the text concerned with powder comminution [5].

A transformation from elastic fracture to plastic deformation will occur when $F_{crack} = F_{yield}$, and so we can write

$$(1 - w/d)w = [2E\gamma_\pi d/3(\sigma_y)^2]^{\frac{1}{2}} \tag{1.24}$$

Equation (1.24) enables us to see that different shapes, that is different combinations of w/d, will produce different transition sizes for the process of comminution in the mill, rather than just deformation and shape change of the particles. The dimension w depends upon the force with which a particle was hit or squeezed in the mill and the plastic flow properties of the material; thus it can be eliminated from equation (1.22) by using equation (1.24) and F_{crack} is then determined by a quadratic:

$$(F_{crack}/B)^2(1/\sigma_y d) - F_{crack}/B + (0.66E\gamma_\pi d)^{\frac{1}{2}} = 0. \tag{1.25}$$

The solution to equation (1.25) has two extremes: for example, when particles are large, that is, d is large, the first term can be neglected and the axial cracking force depends on $d^{\frac{1}{2}}$. However, as the particle size is reduced, the cracking force rises very rapidly until it must exceed the yield stress, σ_y, for the material being milled, however high this may be. At particle size below the critical size, cracking is impossible, and flow must occur instead of comminution. The critical value of d is obtained from equation (1.25) when the $(b^2 - 4ac)^{\frac{1}{2}}$ term in the root expression is put equal to zero; to do this

$$b = -1/B, \quad a = 1/(B^2\sigma_y d) \text{ and } \quad c = (0.66E\gamma_\pi d)^{\frac{1}{2}}.$$

Hence

$$[1/B^2 - 4/B^2\sigma_y d(0.66E\gamma_\pi d)^{\frac{1}{2}} = 0$$

and

$$d = 10.7E\gamma_\pi/\sigma_y^2. \tag{1.26}$$

Equation (1.26) predicts milled particle sizes for ceramics of around 1 μm, because E is typically around 20 GPa, $\gamma_\pi \cong 0.05 \text{ J m}^{-2}$ and σ_y around 100 MPa.

Particle sizes produced by wear grinding as opposed to the axial cracking mechanism modelled above are empirically found from equation (1.27)

$$d = 60000W_{ab}/H_V \tag{1.27}$$

In this approximation, W_{ab} is the work of adhesion and is related to the specific surface work γ_π and the interfacial energy $^a\gamma_b$ through equation (1.28)

$$W_{ab} = (\gamma_\pi)_a + (\gamma_\pi)_b - {}^a\gamma_b. \tag{1.28}$$

In equation (1.27) H_V is the Vickers hardness of the mechanically weaker component (either a or b). The subscripts a and b refer to powder and to the material from which the balls of the mill are made, respectively.

These restriction led technologists to develop ultrafine powders and fibres for making the new **nanophase ceramics** by routes other than mechanical processing.

Nanophase ceramics are shapes formed from powders with particles < 50 nm in diameter, which is some hundred times smaller than the conventional powders prepared by grinding. Nanophase ceramics are more ductile and more easily formed as the grain size decreases; sintering temperatures can be lowered by 500°C.

Powders of this size are currently made by vaporizing a source material in a gaseous flow followed by condensation.

However, the preparation of synthetic powders from solutions is the most common process, and is often called the wet chemical powder preparation technique. Starting materials are milled, dissolved in acids, water, alkalis or molten salts, often at elevated temperatures and pressures. At this stage actions can be taken to control purity by such methods as electrolysis, precipitation or crystallization. Solutions containing the necessary cations in

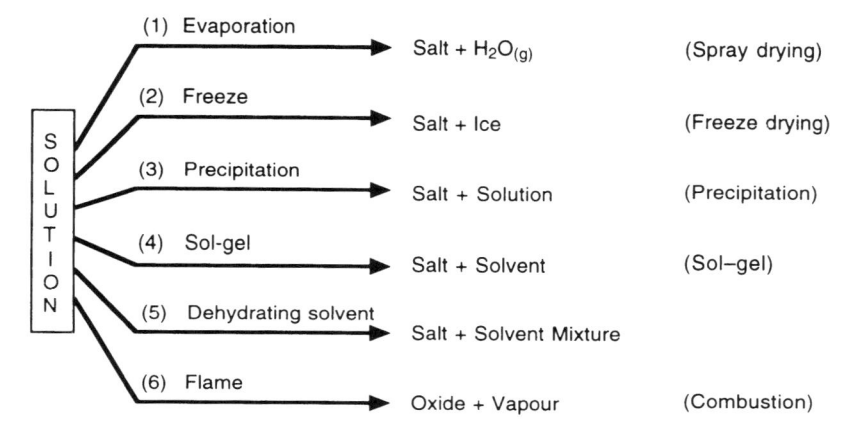

Fig. 1.10 The names given to the variations used in the wet chemical technique for ultrafine powder preparation.

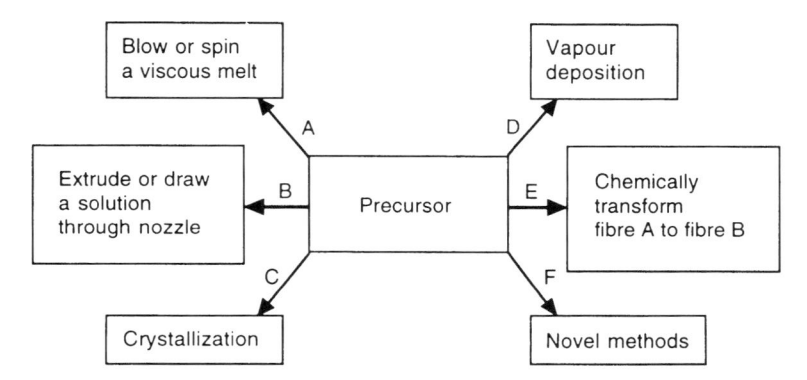

Fig. 1.11 Various processes by which ceramic fibres can be made with dimensions sufficiently small to realize their potential.

Table 1.1 Data relating to the preparation, phase analysis, size and properties of some commercially available ceramic fibres

Company	Name	Phase and preparation method	E (GPa)	σ (GPa)	d (μm)	ρ (kg m^{-3})
Nippon-Carbon	Nicalon	α-SiC + carbon layer polymer degradation	180–210	2.5–3.3	10–20	2550
AUCO	SCS-6	α-SiC + carbon coat CVD[a] on to C	406	3.92	143	3000
Ube	Tyranno	Amorphous SiC 2 wt% Ti retards crystallization polymer degradation	200	2.95	8–10	2400
Dow-Corning	MPDZ	amorphous SiC/Si$_3$N$_4$ polymer degradation	175–210	1.75–2.0	10–15	2350
Dupont	FP	α-Al$_2$O$_3$ sol-gel	385	1.4	20	3900
Sumimoto	—	Al/Si/O$_4$ spinel sol-gel	240	2.2	9–18	3200
3 M	Nextel 312	AlBO$_4$$^+$ mullite sol-gel	154	1.75	11	2700
	Nextel 440	γ-Al$_2$O$_3$$^+$ mullite sol-gel	189	2.10	10	3050
Dupont	PRD-166	α-Al$_2$O$_3$$^+$ PSZ[b]	385	2.35	20	4200

[a]CVD–Chemical vapour deposition
[b]PSZ–Partially stabilized zirconia

the correct ratios are prepared from the purified solutions. Here the skill of the chemist may be paramount, because of solubility problems.

Sulphates and chlorides are the most commonly encountered salts used for solution methods of powder preparation, mainly because the usually soluble nitrates have the drawback that most nitrates melt before decomposition and can produce explosive mixtures when in contact with organic materials.

Variations in the wet chemical powder preparation technique, Fig. 1.10, centre around the way the water or other solvent is removed so that homogeneity is maintained and powder agglomeration is avoided or minimized. Agglomeration prevention is covered in Chapter 3. The calculations in later sections of this chapter indicate that fibres also have to be made with dimensions similar to those of powders if their strength and toughness potential is to be achieved, and here the methods used can be divided into the six groups shown in Fig. 1.11.

At present the application of these methods does not provide fibres with diameters approaching those predicted as being needed, as the data for several commercial ceramic fibres in Table 1.1 show.

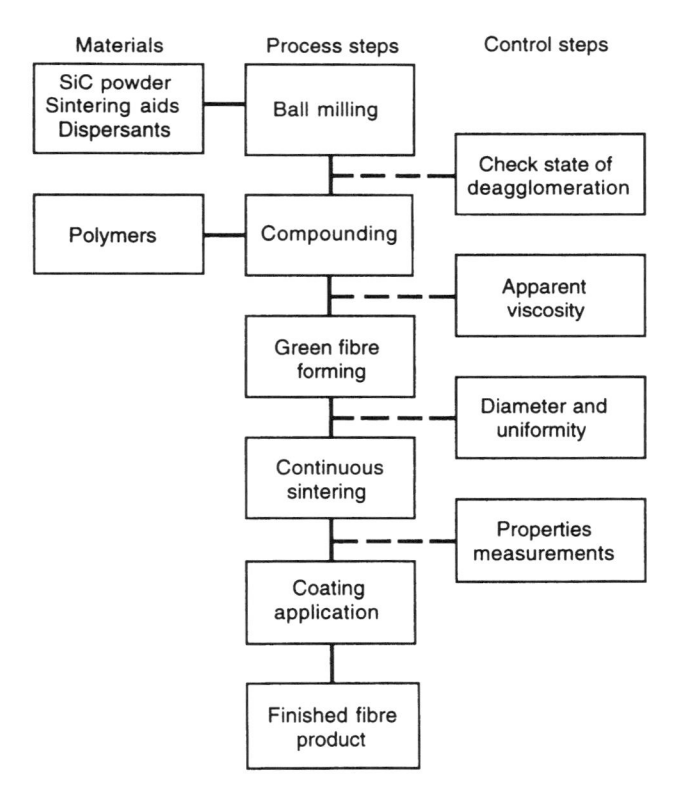

Fig. 1.12 The stages, shown as a flow chart, for a continuous process being developed to produce sintered SiC fibres.

Clearly, vapour phase whisker production is the way forward here to achieve the promise of Fig. 1.8; for SiC this is being developed via the reaction between SiO_2 and carbon.

More commonly at present fibres, rather than whiskers, are produced by semi-continuous processes such as the one, shown as a flow chart, for SiC fibre manufacture in Fig. 1.12.

1.7   COMBINATION OF ULTRAFINE CERAMIC POWDERS AND FIBRES: CERAMIC–CERAMIC COMPOSITES

Previous sections of this chapter were concerned with establishing criteria for high strength in ceramics and in identifying potential ceramics. To date the overriding objective has been to make stronger and stronger ceramics in the sense that higher fracture-initiation stresses are required to bring about failure. However, once the fracture is initiated, there is still minimal resistance to crack propagation in these super-ceramics and spectacular catastrophic failure occurs. Some thought must now be given to developing **tough** ceramics from these strong materials.

Here we are considering toughness as a material property which gives a measure of resistance to crack propagation when that material is subjected to mechanical or thermal stress. One approach to developing this property is to consider the energy demand curve for crack propagation in conjunction with the energy release curve; this statement strongly suggests a Griffith approach.

Griffith's energy balance implies that the energy demand curve will be a straight line

$$U_d = 4c\gamma \tag{1.29}$$

where c = crack length, γ is the surface energy per unit area, and U_d is the energy required to increase the crack length.

The energy release curve will be a parabola of the form

$$U_r = (\pi c^2 \sigma_{app}^2)/E. \tag{1.30}$$

In equation (1.30) U_r is the energy released by the crack, σ_{app} is the applied stress and E is Young's modulus. Catastrophic crack propagation occurs when the slopes of the two curves expressed by equations (1.29) and (1.30) are equal; that is when

$$dU_r/dc = dU_d/dc. \tag{1.31}$$

We can differentiate (1.29) and (1.30) and in accord with equation (1.31) obtain

$$2\pi\sigma_{app}c/E = 4\gamma. \tag{1.32}$$

Since in a pure homogeneous material, on this model, the energy demand is constant, but the slope of the parabola increases, then once the crack is initiated, equation (1.31) must be fulfilled, and instantaneous failure results. To develop a tough ceramic where stable crack propagation occurs, it follows that the energy demand curve must be made non-linear and become concave-upwards so that the gradient of the parabolic curve does not 'catch-up' with the demand curve, or at least higher applied stress levels will be needed to make it do so. This situation is impossible for single-phase, homogeneous, brittle materials however strong they may be, but can be achieved with multiphase, brittle ceramics in the following way.

In a multiphase ceramic, a crack in the weaker component will propagate, but it may encounter the stronger component and then either cut across this phase or detour around it. Either course means that the energy demand will increase as a step function, and crack propagation will stop, unless the applied stress is appropriately increased. A whole series of such encounters produces a series of steps in energy demand and the linear demand curve is changed, becoming in the limit curved. If the second phase is stronger than the matrix phase, then the energy demand curve shape is the desired concave-up, but the reverse is true for a weak second phase in a strong matrix.

Thus, in a multiphase ceramic, as shown schematically in Fig. 1.13, an initial crack of length c_0 will begin to grow under a stress σ_0, but it will be stopped at length c_1 when the slope of the energy demand curve at c_1 exceeds the slope of the energy release curve. For the crack to grow further, the stress has to be raised to σ_1, when the crack will extend from c_1 to c_2 at which point the demand slope curve again exceeds the slope of the release curve. Repeated application of this step mechanism, by which the crack proceeds from instability (moving), to stability (static), eventually produces a situation where the energy release curve cuts the energy demand curve. At the point of intersection, the slopes are equal and the crack will grow spontaneously and

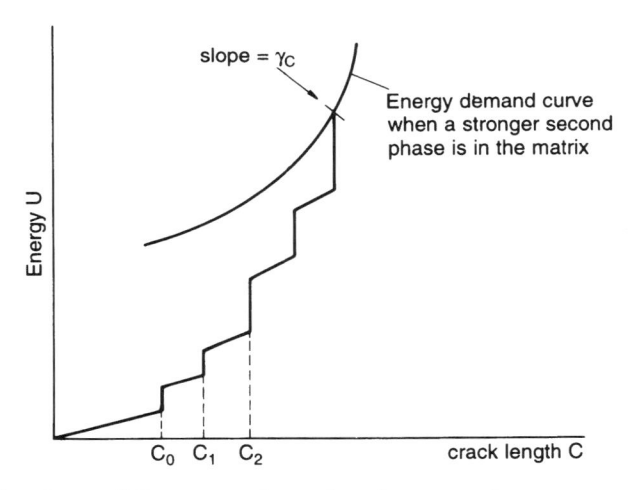

Fig. 1.13 The shape of the energy demand and energy release curves as stated in equations (1.30) and (1.32) for a crack to move through a solid. From [6].

catastrophically. It is the value of the rate of energy absorption at the point of crack instability, i.e. at the point of intersection of the two curves, that is important when considering the toughness of a multiphase ceramic. This value is designated as γ_c, the critical strain energy release rate, and has the value of the maximum slope attained by the energy demand curve. γ_c can be regarded as a material constant which depends on an interplay of factors:

- the type of second phase (tougher or less tough than the matrix);
- the size of second phase particles (too small particles become 'invisible' to the crack, too large ones may depend too much on the interfacial bonds);
- volume of the second phase (increase here results in an increase in γ_c until it becomes large enough to produce microscopic defects).

It is essential in this discussion to realize how important the interphase bonding is. It must be very strong in order to gain the toughness without losing the high strength. Weak interfacial bonding will increase the energy demand, i.e. toughness, by the definition we are using, as the crack detours, but the crack is easier to initiate and ultimate strength is therefore reduced. The inescapable conclusion of this section is that the super-strength ceramics so far discussed will usually provide attractive engineering materials if they are dispersed as a second phase in a matrix material, as long as the interphase bonding has the same order of strength as that present in the strong phase.

So far we have suggested that the best form of the strong materials will be either thin fibres or short whisker crystals. It is obvious to see that such a form will restrict the effect of the strength-reducing flaws. Fine fibres have a small surface, and if this is protected during manufacture then the probability of existence of a strength-damaging flaw becomes very small. Whisker crystals have no dislocations and so have high strength, and furthermore, this type of strength degrader is absent if the strong material has no crystal structure, i.e. it is amorphous.

Thus, the development of stiff, strong and tough engineering ceramics probably lies in developing fibre composites containing the materials described in this chapter. Some considerations applicable to such a development now follow.

1.7.1 Composite toughness

At the beginning of this section the requirements for improving ceramic toughness were discussed, and the conclusion was reached that formation of a multiphase material would improve toughness as long as ideal interphase bonding was achieved. This must apply to systems containing ceramic fibres in ceramic matrices, for a crack generated in the matrix must either detour round the strong fibres (this is called pull-out) or it must break them. Either case, or a combination of both, must increase the energy demand for crack propagation and produce the 'concave-up' curve that is needed to improve toughness.

If there are extremely strong bonds between the fibre and the matrix, it is probable that the major contribution to toughness will be the pull-out work which can be guaranteed in a chopped-strand composite by ensuring that the fibre length does not exceed l_c, the critical length. If l_c is not exceeded then the fibre's ultimate tensile strength will not be reached at any applied stress and cracks will have to detour and cause the pull-out.

It is possible to express this pull-out contribution to toughness in the following manner.

Work to pull out a length of fibre l is given by equation (1.33)

$$\int_0^l \pi d\tau_m x \, dx = d\tau_m l^2/2. \tag{1.33}$$

In equation (1.33) d is the fibre diameter, τ_m the matrix shear stress and l the fibre length. In the composite of fibre volume fraction f, the number of fibres per unit area, N, is given by equation (1.34)

$$N = f/(\pi d^2/4) \tag{1.34}$$

Now assume that pull-out lengths are distributed uniformly between 0 and $l_c/2$ then the total pull-out work is

$$\frac{4f/\pi d^2 \int_0^{l_c/2} \pi d\tau_m(l^2/2) \, dl}{\int_0^{l_c/2} dl} = f\tau_m(l_c)^2/6d, \tag{1.35}$$

and using the equation relating l_c to fibre strength and matrix shear strength $l_c = d\sigma_{fu}/2\tau_m$, the expression for pull-out work becomes

$$\text{pull-out work} = fd(\sigma_{fu})^2/24\tau_m. \tag{1.36}$$

As equation (1.36) shows this can be a large contribution to composite toughness, since σ_{fu} has large values for the materials discussed in this chapter. It is worth noting that thick fibres, where d is large, contribute more to toughness by this mechanism, but strength considerations show that thick fibres will have a lower strength. It follows from this that a combination of long, very thin fibres and short, thick fibres with $l < l_c$, made from the materials described in this paper and embedded in a matrix that produces extremely good matrix–fibre bonds, will produce acceptable properties of high strength with good toughness in a light-weight structure. Perhaps this is the way that these modern ceramics will eventually be developed.

Before the advantages of high strength and improved toughness of wholly ceramic materials can be achieved by utilizing ultrafine ceramic powders and ceramic fibres and/or whiskers, new processing concepts have to be developed in the ceramics industry by learning from related high technology industries. For example, Fig. 1.14 is a flow chart of one such development aimed at improving the process by moving from a batch to a flow process, as used in the polymer industry, and avoiding high temperatures by using the chemical

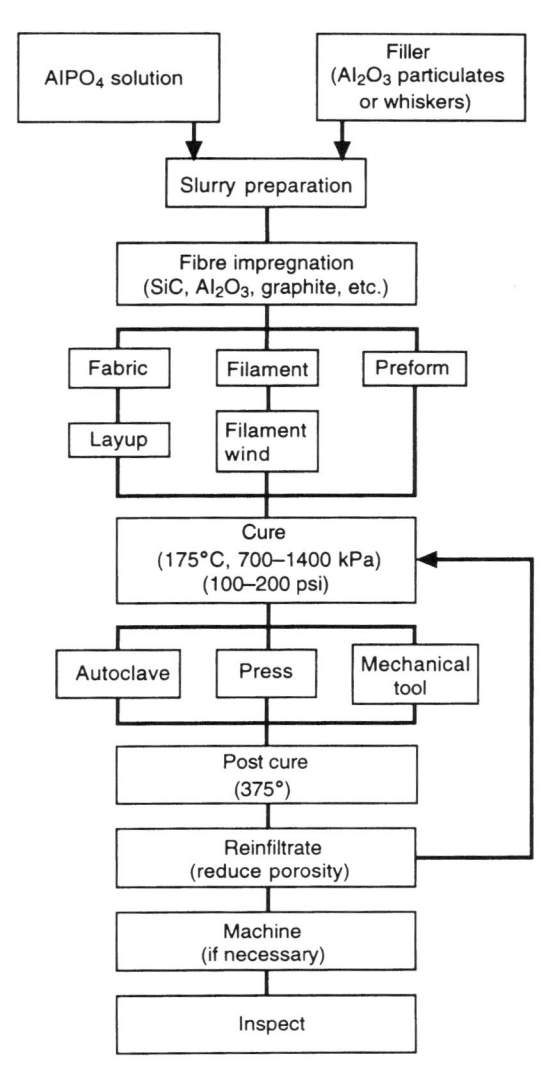

Fig. 1.14 Flow chart of the Alcoa process to make SiC–AlPO$_4$ composites in a continuous process.

bonding potential of phosphate systems. Alcoa are developing this process which can make use of a large range of oxide and non-oxide-coated fibres for reinforcement. The slurry of AlPO$_4$ solution and the whiskers is used to impregnate the chosen fibres which can then be shaped by one of three routes. Curing at 175°C and modest pressure produces a product that is whiskers distributed in berlinite (AlPO$_4$) with no reaction between the infiltrated fibres and matrix. High temperature processing does usually lead to fibre damage.

Clear advantages of the process are

- weight loss less than 1% up to 1650°C,
- linear shrinkage less than 1%,
- use up to 1500°C,
- fracture toughness values up to $16\,\mathrm{MN\,m^{-\frac{3}{2}}}$, and
- tensile strengths up to $312\,\mathrm{MN\,m^{-2}}$.

1.8 RELATIVE COSTS

In order to focus attention on targets with respect to price of these high performance materials the diagram shown as Fig. 1.15 is useful. Before cost-effective ceramics and composites can be achieved, low-cost manufacture of the powders and reinforcements must be achieved unless they are to remain targeted at military, aerospace, prosthetic hip joints and medical implants etc.

Figure 1.15 shows the price/unit weight applied both to materials and products. It contains some measure of value added, perceived values, information content and market size.

Materials are gathered in the left-hand section of the diagram. Industrial sectors are shown on the right-hand side. Those industrial sectors below the £10 kg^{-1} line view materials in a different light to those above this line. Below the line the common feature is that material cost is a major factor of product cost (25% at the line rising to 50% in CIVIL CONSTRUCTION). These sectors seek, from the materials they use, improved properties at no extra cost or improved processability with no loss of performance [4]. Clearly, normal engineering ceramics are currently outside the comprehension of industrial sectors below the £10 kg^{-1} line and so nanophase ceramics, whiskers and composites are irrelevant to their requirements.

Current engineering ceramics are in the 'car-market' so to speak. When nanophase ceramic is combined with carbon fibre, although the product shown in Fig. 1.16, the ceramic–composite knife has vastly improved properties, it is too expensive at £300 per knife.

Can the prices be reduced considerably? Figure 1.17 shows that in the case of whisker manufacture, increase in the pre-conversion materials price increases the whisker cost no matter which of three possible manufacture processes are used. However, if both the production volume and the whisker yield in the modified Acheson process was improved, whisker costs could be brought below £50 kg^{-1} making them candidates for several important industrial sectors. The parallel is with the cost of early silicon chips compared with current prices.

However, before any major improvement is made to the cost/kg of ceramic artefacts a quantum leap has to be made away from the batch process shown in Fig. 1.1 with the high percentage of waste this generates. One potential area that may provide this quantum change is outlined in the final section.

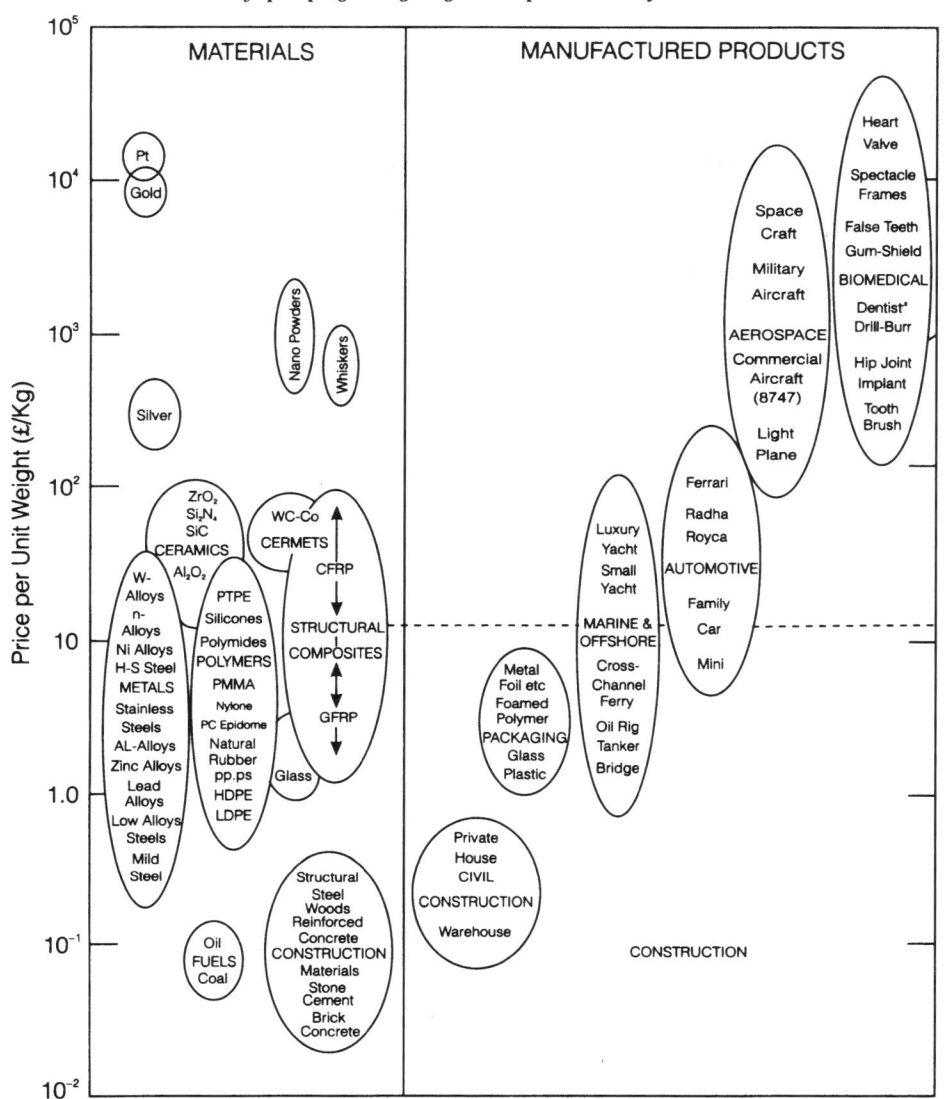

Fig. 1.15 Price per unit weight of materials and products which shows the targets that have to be achieved if widespread use of a particular material is to be achieved. After Ashby [7].

1.9   SELF-PROPAGATING HIGH-TEMPERATURE SYNTHESIS (SHS)

This is a method used to produce both powders and formed shapes; it involves an exothermic reaction producing temperatures above 2500°C.

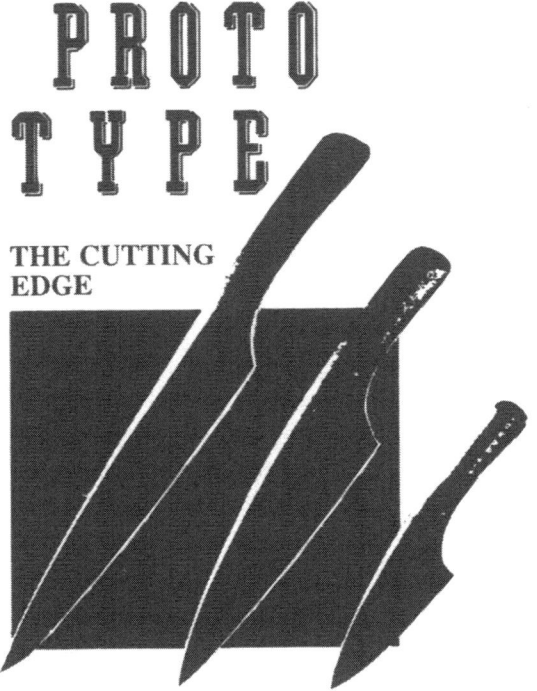

PROTO TYPE

THE CUTTING EDGE

Since the demise of flint as a fashionable cutting material all the best kitchen knives have been made out of steel. This is about to change. Japanese technology has made ceramic the best cutting edge. Richard Seymour of the British design group Seymour Powers started experimenting with the new ceramic after seeing a little pair of ceramic scissors when he was in Japan

"Ceramic has extraordinary potential." he says. Turned into a blade, it will never need sharpening. And for keen cooks like Seymour it offers a new type of cutting aesthetic. "The most delightful thing about the new blade is the sensation." Seymour says."When you finger it, it is warm and fools you into thinking it will be a soft cut. But press gently and it will lase through any kind of food."

As the material is translucent, it also introduces the possibility of colour. The prototype developed by Seymour Powers (see Resources) uses matt black carbon fibre for the handle and top of the blade and fluorescent blue ceramic on the cutting edge. Seymour wanted to suggest the hot/cold sensations of the knife.

A formidable problem at the moment is price – currently £300 a knife. Seymour explains: "The material is made from compressed powder, it is a long-winded process – a little like making silicon chips." But, like silicon chips, it shouldn't be long before the price drops and we will all be using the new knives.

LUCIE YOUNG

Fig. 1.16 A recent advert for a carbon-fibre handled, zirconia-blade, kitchen knife priced at £300.

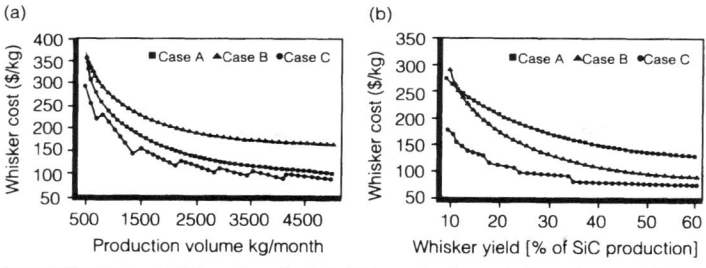

Cost of SiC whiskers could be reduced by both (A) increasing the production volume. and (B) increasing the whisker yield. in a cost analysis of SiC whisker fabrication at IBIS Associates.

Fig. 1.17 The effect on the cost of Sic whiskers of increasing the quantity made and improving the process for three types of manufacturing process (from a cost analysis by IBIS associates).

As long ago as 1966, it was recognized that sintering of ceramic powders could be improved if they were first subjected to an explosive shock treatment. Two contributions to the sintering rate enhancement were recognized

- powder particle size reduction, and
- introduction of a large number of defects during the shock treatment.

Discussion on sintering makes it clear why a shock-treated powder might show improved sintering behaviour from these two effects. The results show that explosive shock treatment of ceramic powders produces increases in both dislocation density (up to 10^{13} cm^{-3}) and in point defect levels (up to 10^{19} cm^{-3}).

These increased defect levels come from the action of planar shock waves travelling through highly compacted powder discs, as well as being generated by the initial explosive charge. Typically, when a powder is subjected to a controlled explosion in a protective copper can, mean pressure of 16–22 GPa and mean temperatures up to 300°C are generated. When an exothermic reaction is also involved, the 300°C is soon raised to nearly 3000°C. Processing times are of the order of seconds instead of hours or days for normal sintering routes.

The high temperature volatilizes most impurities, hence leaving a purer product, and it can produce unusual materials; for example, a cubic form of TaN has been synthesized which is about three times harder than the normal hexagonal form of TaN, which is itself hard ($H_V = 11.1$ GPa). Ceramic composites of the type B_4C–Al_2O_3 and SiC–TiB_2–Al_2O_3 have been made, because the extremely rapid heating rate limits the oxidation of the carbide by the matrix oxide which is an obvious advantage.

For general SHS processes there are two types of reaction.

- Thermite: this usually produces ceramic composites from an oxidation-reduction process.
- Compound formation: in this case elemental powders are used.

Both reaction types have two mechanisms leading to the final product.

Special ceramics for modern applications

(a) Propagating reaction

When a propagating reaction occurs it is initiated locally and spreads out. Thus one conceives of a **synthesis wave** passing through the compacted powder. For the wave to travel smoothly through the mass, the rate of heat dissipation must be less than the rate of heat generation, or the powder is quenched. Heat dissipation is a function of several parameters, each affecting the rate at which the synthesis wave moves through the solid compact. The effect of various parameters is shown schematically in Fig. 1.18.

(b) Bulk reaction

In this reaction, the powder mixture is rapidly heated to cause simultaneous reaction through the whole sample; this can be viewed as a thermal explosion.

In these reactions, particularly in the propagating reaction, an equilibrium condition exists if the process occurs at relatively low temperature or the rate of synthesis wave propagation is low. Under controlled propagation the heat release time zone merges with the reaction time zone and with the structure formation time zone, to produce the zone of synthesis where chemical bonds and microstructure evolve simultaneously. Hence, efforts are most often devoted to control the synthesis by applying a slowing-down of the synthesis wave. The most effective way of achieving this is to add up to 60% of previously reacted product to the starting powders. Occasionally activators are added in an attempt to keep the SHS parameters within the limits shown in Table 1.2.

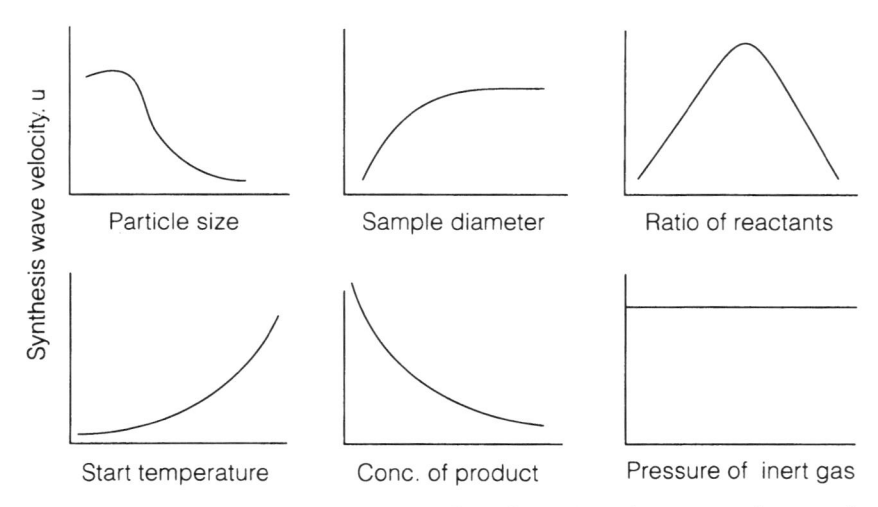

Fig. 1.18 Schematic representation of the effect of sample and process variants on the rate a synthesis wave travels through a compact in the self-propagating high-temperature synthesis method.

Table 1.2 Self-propagating high-temperature synthesis parameter ranges

Parameter	Range
Maximum temperature	$1500-4000\,°C$
Synthesis wave rate	$0.01-0.15\,m\,s^{-1}$
Rate of heating	$10^3-10^6\,°C\,s^{-2}$
Intensity of initiation	$1-10\,kJ\,m^{-1}\,s^{-1}$
Duration of initiation	$0.05-4.0\,s$
Size of synthesis zone	$1 \times 10^{-4}-5 \times 10^{-3}\,m$

When control is achieved, and equilibrium conditions exist, the combustion front moves at a constant linear rate, and this is called stationary combustion. Unstable combustion can arise from the following causes.

- In solid–gas systems, repeated synthesis waves can occur because of poor gas supply to the reaction site. The first synthesis wave is followed by a slower one. This is often the case in synthesis of ceramic nitrides from metal powders in a nitrogen atmosphere, when the adiabatic temperature $T_{ad} > T_{mpt}$ and the porous powder compact is blocked off by liquid product at the surface. Additive control can help in this case, for example if sodium azide (NaN_3) powder is added to the metal powder, then the combustion wave causes decomposition locally to provide N_2 for reaction, while at the same time volatilizing sodium metal,

$$3Ti_{(s)} + NaN_{3(s)} \rightarrow 3TiN_{(s)} + Na_{(vap)}.$$

- The combustion wave can reflect back from the sample surface and interact with the slower-moving secondary wave
- Inclusion of inert components causes the wave front to break up, then move at a variety of rates in various directions.

Unstable combustion leads to an oscillating combustion front which is an oscillating synthesis wave, giving the final product a laminated structure that can sometimes be desirable.

As described, the SHS method is just a rapid way of producing ceramic powders, and is used as such surprisingly widely in Russia, where over 1000 tons per annum each of $MoSi_2$, Si_3N_4, TiC and an $Al_2O_3-B_4C$ composite are made. Effort is now directed towards simultaneous densification of the product, still within the short reaction time scale of SHS. The first rule of densification methods is that densification attempts must be synchronized with the passage of the synthesis wave: therein lies the difficulty. Attempts to apply pressure to the system before this, in say the heating initiation time zone, prevent volatiles such as adsorbed H_2O from escaping. This leads to die rupture or to specimen bloating. Figure 1.19 shows two types of equipment developed in Australia and

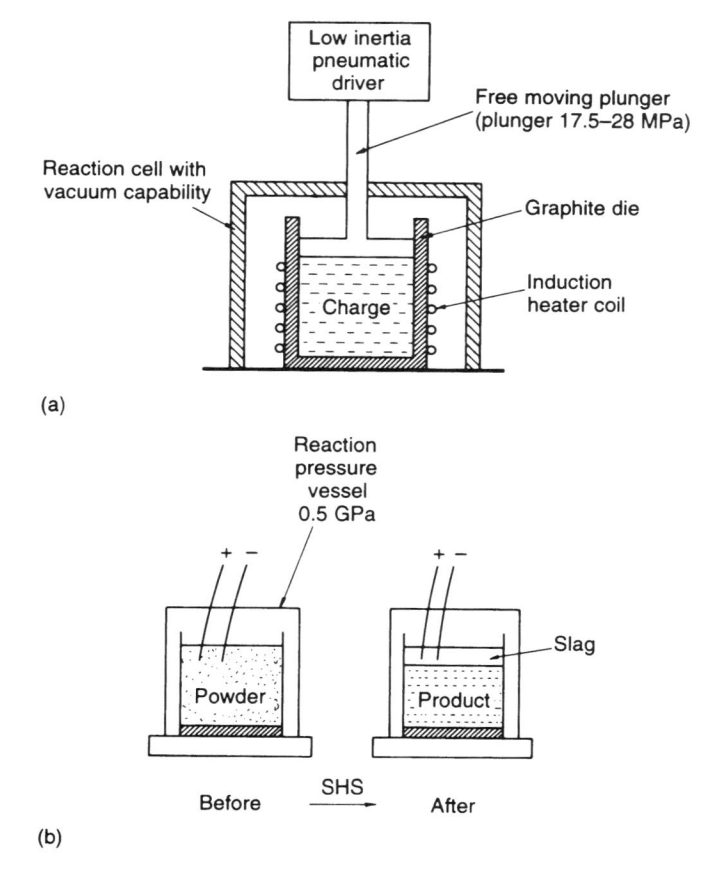

(a)

(b)

Fig. 1.19 Apparatus designed to apply pressure to a sample in the self propagating hight temperature synthesis method as the synthesis wave moves through the sample.

in Russia to apply suitable pressures as the synthesis wave moves through the reactive charge. More sophisticated equipment is based on a sequential application of vacuum out-gassing, followed by rapid application of high gas pressure by a gas gun technique, so that swelling is constrained and densities in the 90–95% theoretical range are achieved within seconds.

To obtain densities in excess of 99% after the sequential operation of vacuum and gas gun, mechanical deformation must be applied while the product is still hot after the passage of the synthesis wave. Following the synthesis wave in steady-state combusion by a hot rolling procedure can produce theoretical density, but has the inherent problem of matching the velocity and position of the rolls with the velocity and position of the wave front.

However, if these difficulties can be overcome then an extremely rapid process could be developed, able to produce the microstructure shown as desirable in Fig. 1.1 at a fraction of the current costs.

REFERENCES

1. Ashby, M.F. and Jones, D.R.H. (1982) *Engineering Materials: An Introduction to Their Properties and Applications*, Pergamon Press, Oxford.
2. Griffith, A.A. (1920) *Phil. Trans. R. Soc. Lond. A*, **221**, 163.
3. Orowan, E. (1945) *Proc. Inst. Eng. and Shipbuilders in Scotland*, **89**, 165.
4. McColm, I.J. and Clark, N.J. (1988) *The Forming, Shaping and Working of High-Performance Ceramics*, Blackie and Sons, Glasgow.
5. McColm, I.J. (1990) *Ceramic Hardness*, Plenum, New York.
6. McColm, I.J. (1983) *Ceramic Science for Materials Technologists*, Leonard Hill, Glasgow.
7. Ashby, M.F. (1991) The price–unit weight diagram, in *Industrial Materials for the Future* DTI report (eds D.A. Melford, P.D.R. Rice, W. Bonfield, N.A. Waterman and M.F. Ashby), HMSO, London, pp. 7–10.

CHAPTER 2

Laser vapour phase synthesis of ceramic powders

R.A. Bauer and J. Schoonman

2.1 INTRODUCTION

2.1.1 General introduction

Recent interest in high temperature structural ceramics has led to the development of chemical vapour phase synthesis techniques for the formation of a variety of high quality oxidic and non-oxidic ceramic powders. Silicon nitride, Si_3N_4, is one of the most promising of these structural ceramics, because of its high-temperature strength, thermal shock resistance and corrosion resistance. However, the absence of a liquid phase of Si_3N_4 in the sintering process hinders densification, resulting in a less optimal microstructure after densification and hence lowering the strength of the formed ceramics. Therefore, sinter additives and hot isostatic pressing (HIP) are commonly employed to improve the densification. These sinter additives also reduce high-temperature performance because of liquid phase formation, while HIP increases the cost significantly.

By controlling the particle packing in the green forms, the microstructure and properties can be controlled better. By preparing green densities from small, equiaxed, non-agglomerated particles optimal green densities can be achieved [1]. Additionally, reducing the particle size introduces enlarged surface areas and energies and, therefore, increases the driving force for densification. This allows the use of more economical sintering techniques. Besides, chemical impurities in the starting powders may exert a negative influence on densification and have to be avoided. In addition to the above mentioned prerequisites for obtaining dense ceramics, the starting powder should be very pure. This chapter describes the results of an investigation of highly sinteractive powders of Si and Si_3N_4 by gas phase deposition techniques.

Ceramic Processing. Edited by R.A. Terpstra, P.P.A.C. Pex and A.H. de Vries.
Published in 1995 by Chapman & Hall, London. ISBN 0 412 59830 2

Si_3N_4 powders have been prepared by several techniques, such as direct nitridation of silicon, decomposition of silicon diimide and vapour phase synthesis. Although relatively small particles can be produced by using vapour phase synthesis initiated by convective heat [2], smaller and especially less agglomerated powders have been prepared using a sophisticated technique, laser-chemical vapour precipitation (L-CVP), developed by Haggerty [3–6] and Danforth [3, 4, 7]. Characteristic of this laser-driven gas phase synthesis process is a well defined reaction zone which facilitates a considerable degree of control over composition, size and size distribution of the product powders. In the formation of silicon (Si), silicon carbide (SiC) and silicon nitride (Si_3N_4) by L-CVP, silane (SiH_4) is invariably used as the silicon reactant [3–7]. However, with the use of SiH_4 it is difficult to obtain a stoichiometric product in the case of SiC and Si_3N_4. Often an excess of silicon has been observed [4]. Both thermodynamic calculations and experimental data show that less silicon contamination is likely to occur with increasing halogen content of the system [8]. Whereas in L-CVP typically SiH_4 is used, in the traditional thermally activated CVP of powders and thin films of Si_3N_4 chlorinated silanes, i.e. dichlorosilane (SiH_2Cl_2), trichlorosilane ($SiHCl_3$) and silicon tetrachloride ($SiCl_4$), are frequently employed [8].

Recent work [2] indicates that the synthesis of spherical, submicron powders of amorphous Si_3N_4 with a narrow size distribution is feasible using chlorinated silanes in thermally activated CVP. This research has been extended to the use of chlorinated silanes in the L-CVP of Si and Si_3N_4 powders. Silicon can be formed from SiH_2Cl_2 according to the reaction

$$SiH_2Cl_2 \rightleftarrows Si_{(s)} + 2HCl. \tag{2.1}$$

The addition of ammonia (NH_3) to laser excited SiH_2Cl_2 results in the formation of Si_3N_4.

$$3SiH_2Cl_2 + 4NH_3 \rightarrow Si_3N_{4(s)} + 6HCl + 6H_2. \tag{2.2}$$

A similar reaction scheme for the production of Si_3N_4 can be given for other chlorinated silanes as silicon source, viz. $SiHCl_3$ and $SiCl_4$. Chlorinated silanes react readily with NH_3 to form silicon diimides at ambient temperatures [9–11].

$$SiH_2Cl_2 + 2NH_3 \rightarrow Si(NH)_{2(s)} + 2HCl + 2H_2. \tag{2.3}$$

Whereas the $Si-H_2$ bond vibration in SiH_4 absorbs the radiation of a CO_2-laser (10.6 μm) very efficiently, chlorinated silanes and in particular $SiHCl_3$ and $SiCl_4$, can be expected to show considerably less absorption of the CO_2-laser spectrum. Hence, it is important to study other potential energy transfer routes such as by indirect excitation using sensitizers.

2.1.2 The use of sensitizers

The first law of photochemistry stating that 'only light that is absorbed by a molecule can be effective in producing photochemical change in a molecule' was expressed by Calvert and Pitts [12]. If a molecule does not absorb a sufficient proportion of the incident light to be brought to the desired excited state, the utilization of a sensitizer is essential. Ideally, a sensitizer strongly absorbs radiation and transfers a sufficient portion of the absorbed energy to the gaseous reactant species without decomposing or participating in the reaction sequence:

$$S \xrightarrow{\text{IR } h\nu} S^{**}$$

$$S^{**} + R \overset{\text{collisions}}{\rightleftharpoons} S^* + R^* \tag{2.4}$$

$$R^* \longrightarrow P$$

In visible and ultraviolet photochemistry sensitizers, such as Hg, are commonly used. With the availability of CO_2-lasers the possibility of inducing chemical reactions, especially organic reactions, by infrared irradiation was successfully explored in the 1970s. Interest was aroused in the use of sensitizers for the $9-10 \, \mu m$ emission region. Shaub and Bauer [13] pioneered infrared sensitizers in a process, which they referred to as laser powered homogeneous pyrolysis. They utilized SF_6 as a sensitizer for radiation from a CW (continuous wave) CO_2-laser to heat organic molecules to temperatures between 500 and 1500 K.

Normally, untuned CO_2-lasers are used which emit only the P20 laser line of the $00°1-10°0$ band at $10.4 \, \mu m$. Therefore, SF_6 is often preferred as sensitizer, especially as an energy transfer agent for organic molecules [13, 14]. SF_6 has an extremely high absorption coefficient in the $10 \, \mu m$ region of the infrared spectrum. Its absorption coefficient at the P20 line of the $00°1-10°0$ band at $10.4 \, \mu m$ has been measured by several authors, i.e. $350 \times 10^{-3} \, Pa^{-1} \, m^{-1}$ according to Brunet [15] at low pressure ($10^{-3} \, Pa^{-1} \, m^{-1} = 1.013 \, atm^{-1} \, cm^{-1}$), and $400 \times 10^{-3} \, Pa^{-1} \, m^{-1}$ at 500 hPa according to Nowak and Lyman [16]. The absorption coefficient of one of the strong absorbing chlorinated silane reactants, SiH_2Cl_2, used in this work is about 500 times lower at the same line, i.e. $0.85 \times 10^{-3} \, Pa^{-1} \, m^{-1}$ as measured at 5 hPa [17]. However, the use of SF_6 is limited to temperatures below 1600 K and to hydrogen-poor conditions, since above these temperatures SF_6 dissociates especially in the presence of hydrogen [13, 18, 19]. Bauer and Haberman [20] have shown that SF_6 even reacts explosively with SiH_4, when irradiated by a CO_2-laser. SF_6 also reveals strong saturation when high radiation densities are used [21]. Therefore, the absorption coefficient of SF_6 for high laser powers will be considerably lower.

In laser chemistry, interest was also aroused in the use of SiF_4 as a sensitizer [22] because of its strong absorption of CO_2-laser radiation and its excellent stability compared to other IR-absorbing sensitizers, e.g. SF_6. Its inertness can be explained by the high bonding enthalpy, ΔH_b, of the Si–F bond, i.e. $596 \, kJ \, mol^{-1}$ at room temperature [23] as compared with other Si bonding enthalpies, e.g. ΔH_b (Si–H) $= 322 \, kJ \, mol^{-1}$. On the other hand, the average bonding enthalpy of the S–F bond in SF_6, $329 \, kJ \, mol^{-1}$, is considerably weaker, also compared to other S bonding enthalpies, e.g. ΔH_b (S–H) $=$

$367 \, \text{kJ} \, \text{mol}^{-1}$. The triply degenerate Si–F stretching mode, v_3, of SiF_4 is centred around $1032 \, \text{cm}^{-1}$ (9.69 μm), and has an exceptionally high Beer's law extinction coefficient of $91 \times 10^{-3} \, \text{Pa}^{-1} \, \text{m}^{-1}$, as reported by Olszyna *et al.* [22]. This absorption peak is close to the P42 line of the $00°1–02°0$ band of the CO_2-laser, which is centred at $1025 \, \text{cm}^{-1}$ (9.76 μm).

The success of the L-CVP strongly depends upon the choice of reactants. Most of the reported research has dealt with the production of silicon nitride, Si_3N_4, from the reaction of silane, SiH_4, with ammonia, NH_3. The silane strongly absorbs the radiation emitted by a CO_2-laser which has a wavelength of 10.6 μm. In this process, however, it seems to be difficult to obtain a stoichiometric product, an excess of silicon has often been reported. For this work the investigations were based on the utilization of chlorinated silanes, SiH_2Cl_2, $SiHCl_3$ and $SiCl_4$, because they are commercially more attractive and safer. However, the latter two reactants have a negligible absorption coefficient of CO_2-laser radiation whereas SiH_2Cl_2 shows a broad absorption feature in the 10.4 μm region. In this chapter the use of the sensitizers, SF_6 and SiF_4, is explored to excite the chlorinated silanes, $SiCl_4$ and $SiHCl_3$ by a CO_2-laser, and to initiate the formation of Si_3N_4 powders from these two chlorinated silanes and NH_3. In addition, the effect of the sensitizers on the excitation of SiH_2Cl_2 is compared to the primary excitation of SiH_2Cl_2.

2.2 EXPERIMENTAL ASPECTS

The reactant gases and reactor used are the same as reported before [6]. A 200 W CW CO_2-laser (PL6, Edinburgh Instruments Ltd., Edinburgh, UK) with a wavelength tuneable between 9.166 μm and 10.886 μm was partially focused with a ZnSe lens (focal length 30 cm) to a spot of 1.5 mm in diameter in an orthogonal stream of the reactants.

The gas inlet nozzle consisted of four concentric inlets ($d = 0.4 \, \text{mm}$) for NH_3 pointed towards the middle, and a central outlet ($d = 1 \, \text{mm}$) for the chlorinated silane. The gases were mixed in the laser beam in order to avoid low temperature reactions.

In all the experiments, when SF_6 was employed as sensitizer, the P20 laser line $00°1–10°0$ band at 10.4 μm, also referred to as the 10P20 line, was used to irradiate the reactant gases with a laser power of 130 W. On the other hand, for all the gas mixtures with SiF_4 as sensitizer, the P42 laser line of the $00°1–02°0$ band at 9.4 μm, was utilized to heat this gas mixture with a laser power of 90 W.

$SiHCl_3$ and SiH_2Cl_2 were transferred directly into the reactor but the liquid reactant $SiCl_4$ (b.p. = 330 K) was first evaporated into a reflux condenser. The flow and partial pressure of $SiCl_4$ was controlled by the flow of the carrier gas, N_2, through the condenser and by the condenser temperature. The condenser temperature was regulated at 311 K with a constant temperature bath (Tamson, TLC 3) such that the flow rate of $SiCl_4$ equalled the flow rate of the carrier gas.

It was not possible to use a brightness pyrometer to measure the temperature, because of the high transmission of the flame. To measure the flame temperature, a two-colour pyrometer in combination with a spectrometer was used, as described before [17].

The thermodynamic equilibrium compositions of input mixtures of SiF_4 and NH_3 were calculated using a computer program SOLGASMIX [24], a procedure originally developed by White *et al.* [25]. It calculates the composition which has the minimum free energy for the system under given conditions of temperature, total pressure and input gas concentrations. The following 83 species have been included for the thermodynamic calculations: $SiCl_4$, $SiHCl_3$, SiH_2Cl_2, $SiCl_2$, HCl, H_2, Cl_2, $SiCl$, $SiCl_3$, SiH_3Cl, SiH_4, Cl, H, SiH, Si, Si_2, Si_3, N, N_2, N_3, NH, NH_2, NH_3, SiN, Si_2N, $N_2H_2(cis)$, N_2H_4, F_7H_7, F_6H_6, F_5H_5, SiF_4, N_2F_4, F_4H_4, SiF_3, NF_3, F_3H_3, $SiHF_3$, SiF_2, $N_2F_2(cis)$, $N_2F_2(trans)$, NF_2, SiH_2F_2, F_2H_2, F_2, SiF, NF, SiH_3F, F, HF, $SiCl_3F$, ClF_5, $SiClF_3$, ClF_3, ClF, SF, SF_2, FS_2F, S_2F_2, SF_3, SF_4, SF_5, SF_6, S_2F_{10}, HS, H_2S, Si_2H_6, S, SSi, S_2, S_3, S_4, S_5, S_6, S_7, S_8, $S_{(1)}$, $S_2Si_{(s)}$, $S_2Si_{(1)}$, $NH_4Cl_{(s)}$, $N_2H_{4(1)}$, $Si_{(1)}$, $Si_{(s)}$, $Si_3N_{4(s)}$. The necessary thermodynamic values, the enthalpies, entropies, and heat capacities for all species were taken from JANAF [23]. Unfortunately, there were no quantitative thermodynamic data available for both $(SiCl_2)_n$ [26, 27] and for $Si(NH)_2$ [9, 10, 11] and other imide-like species, although their occurrence especially at temperatures below 1200 K is likely, respectively, for nitrogen-poor and nitrogen-rich systems. Therefore, at moderate temperatures, in both nitrogen-poor and nitrogen rich systems the corresponding yield of Si and Si_3N_4 is expected to be lower than calculated.

The powders were collected in an electrostatic precipitator [6] at a temperature of 570 K. Afterwards the powder collected was heated to 770 K in vacuum for 15 h in order to remove the remaining traces of NH_4Cl. The powders were analysed using IR-spectroscopy and characterized by X-ray diffraction (XRD) and electron diffraction on the transmission electron microscope (TEM). The amounts of nitrogen in the powders were determined by decomposing the samples in LiOH at 950 K and measuring conductometrically the amount of NH_3 released [28]. The relative weights of chlorine and of sulphur in the powders were determined by means of microtitration with $AgNO_3$, and by means of inductively coupled plasma (ICP) analysis respectively. Particle sizes were measured from TEM micrographs. The specific surface area of the powders was determined using gas adsorption measurements (BET).

2.3 RESULTS

2.3.1 Optical absorption measurements

The absorption coefficient of SiH_2Cl_2 was measured and found to be $0.85 \times 10^{-3}\,Pa^{-1}\,m^{-1}$ at the P20 CO_2 line of the $00°1$–$10°0$ band at $10.4\,\mu m$

and at a partial pressure of 50 hPa. The absorption of CO_2-laser radiation by the other two chlorinated silanes used, $SiHCl_3$ and $SiCl_4$, is negligible. The use of NH_3 as absorbing reactant reveals little absorption of the CO_2-laser lines, P42 of $00°1-02°0$ band at 9.4 µm and the P20 of $00°1-10°0$ band at 10.4 µm. Therefore, the effect of NH_3 on the asorption of CO_2-laser radiation can be neglected in the present study.

The absorption of the CO_2-laser lines by both SF_6 and SiF_4 was measured in a 10 cm cell [17]. The absorption coefficient was calculated according to the Lambert–Beer law:

$$\alpha = -\frac{\ln\left(\frac{I_t}{I_0}\right)}{LP_s} \qquad (2.5)$$

where α denotes the absorption coefficient, L the path-length of the cell, P_s the partial pressure of the sensitizer, and I_0 and I_t the incident and transmitted intensity of the CO_2-laser, respectively.

The cell contained one of the sensitizers with a partial pressure of 5 hPa and was filled to atmospheric pressure with N_2. It was not possible to measure absorption values at 10^5 Pa because at this high pressure the laser radiation is almost completely absorbed by the sensitizer. The absorption coefficients of SF_6 at the P20 CO_2 line of the $00°1-10°0$ band at 10.4 µm, also referred to as 10P20, and of SiF_4 at the P42 CO_2 line of the $00°1-02°0$ band at 9.4 µm, also referred to as 9P42, were measured to be 30×10^{-3} and 123×10^{-3} $Pa^{-1}\,m^{-1}$, respectively, at a partial pressure of 5 hPa. While the absorption coefficient of SiF_4 is comparable to the literature values, the measured absorption coefficient of SF_6 is much smaller than the literature value of about 0.4 $Pa^{-1}\,m^{-1}$ [15, 16]. During absorption measurements, the absorption varied little with time, which indicates that the relatively low absorption values is not caused by dissociation of SF_6. It can be explained by a phenomenon referred to as 'hole burning' [21], where local saturation of rotational fine structure levels in a molecule prevents further excitation and thus lowers the absorption of the molecule. Brunet [15] showed that SF_6 easily saturates using moderate power densities.

The absorption coefficient varies with total pressure, partial pressure of the absorbing species and laser intensity. Therefore, it is necessary to measure the effect of the sensitizer used and of the concentration of the sensitizer under experimental conditions. Here a two-colour pyrometer was employed to measure the temperatures of condensed phases in the gas mixtures. It was shown before, that by heating SiH_2Cl_2 with CO_2-laser radiation, silicon particles are formed. These particles are either liquid above a temperature of 1683 K, or solid below this temperature. Therefore, by heating a mixture of SiH_2Cl_2 and SiF_4 or SF_6, the temperature as a function of the amount of sensitizer can be determined. While two-colour pyrometry was used to measure the temperature, it has to be verified that the emission originated from a black or grey body. The visible spectra of the mixtures used, with SiH_2Cl_2 as

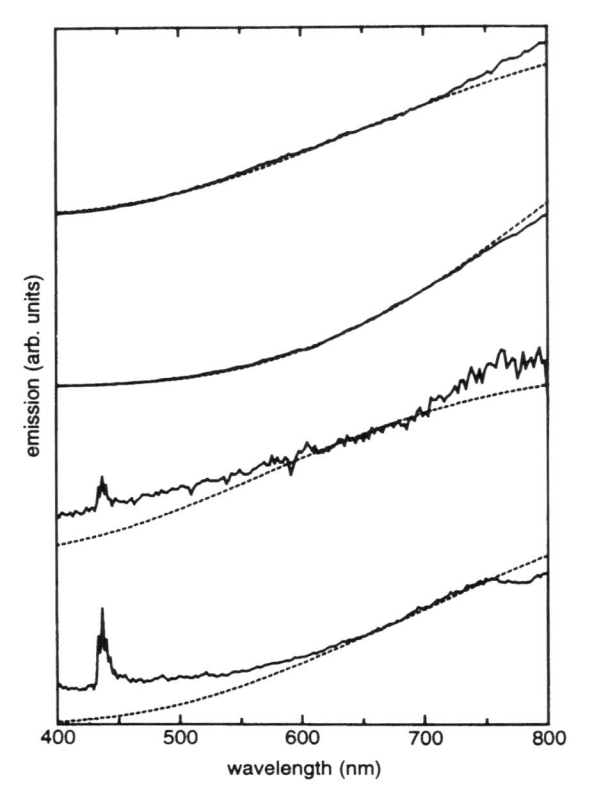

Fig. 2.1 The visible spectra along with the fitted black body curves, i.e. the dotted lines, of laser heated gas mixtures containing SiH_2Cl_2. From top to bottom: $SiF_4-SiH_2Cl_2-NH_3$, $SiF_4-SiH_2Cl_2$, $SF_6-SiH_2Cl_2-NH_3$, $SF_6-SiH_2Cl_2$.

silicon source, and their black body fits are shown in Fig. 2.1. These spectra reveal that the laser flames, with SiF_4 as sensitizer, show almost perfect black body spectra. On the other hand, the spectra containing SF_6 do not compare so well with black body curves. Figure 2.2 shows the temperature attained, as measured with two-colour pyrometry for SiH_2Cl_2 and a variable amount of sensitizer. In the experiment with SF_6 the flow rate of SiH_2Cl_2 was kept at 53 standard cubic centimetres per minute (sccm) while varying the sensitizer flow rate. Because SiH_2Cl_2 also absorbs the laser line P20 of the 10.4 μm band used for SF_6, the gas flow is also heated without a sensitizer. In the other experiments, where SiF_4 was used as sensitizer, another laser line, P42 line of the 9.4 μm band was used to excite the sensitizer. SiH_2Cl_2 did not absorb this later laser line. Since the absorption of SiF_4 is weaker compared to SF_6, the flow rate of the sensitizer had to be varied over a larger range. Therefore, to limit disturbance from differences in gas flow velocity, the total flow was kept at a constant value of 70 sccm, while varying both SiH_2Cl_2 and SiF_4 flow rates.

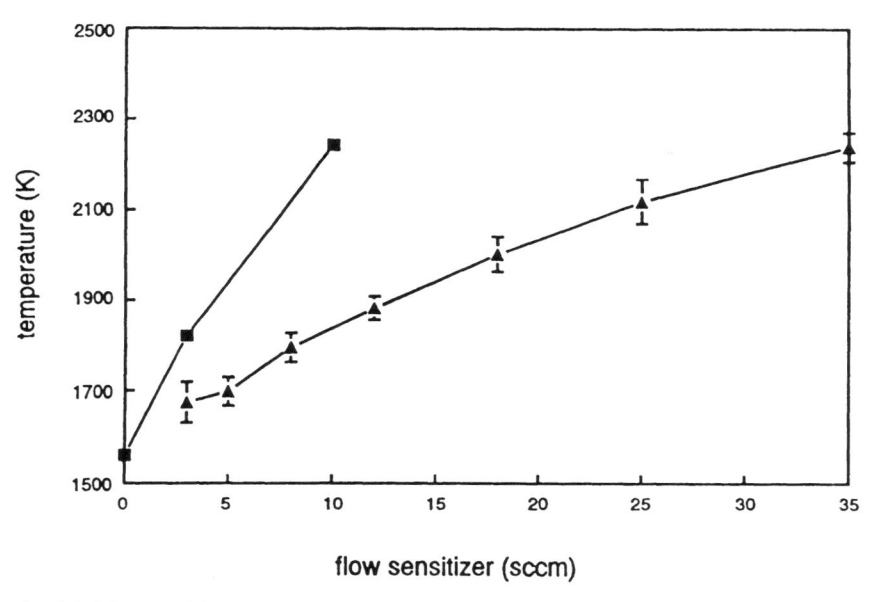

flow sensitizer (sccm)

Fig. 2.2 The resulting temperature of silicon particles formed from SiH_2Cl_2 while varying the flow of the two sensitizers: ■ = SF_6, ▲ = SiF_4.

The heat release from the flame increases with increase of the reaction temperature. Notably, because of the high particle density, the heat loss by radiation from the particles formed in the flame plays an important role in the reduction of the temperature. Thermal radiation is very sensitive to temperature, it increases with the fourth power according to the Stefan–Boltzmann law.

The results in Fig. 2.2 reveal that the temperature increase as a function of the sensitizer flow rate is larger when SF_6 is used. Therefore, it is apparent that the absorption of SF_6 is stronger than the absorption of SiF_4.

2.3.2 Thermodynamic calculations

Gas phase nucleation reactions are very complex with many parameters requiring optimization in practice. Thermodynamic calculations can be used to assist in the evaluation of a reaction system, to optimize conditions, and to gain an understanding of the system. By thermodynamic calculations, deposition phase fields and theoretical deposition efficiencies of condensed phases can be calculated under thermodynamic equilibrium. The efficiencies are expressed as a percentage of the maximal conversion of the source materials. Kinetic limitations will only further reduce the practical efficiency.

Deposition efficiencies at atmospheric pressure of silicon from SiH_2Cl_2, $SiHCl_3$, $SiCl_4$ and SiH_4, were calculated using SOLGASMIX. The results are shown in Fig. 2.3. Whereas SiH_4 is virtually completely converted into silicon,

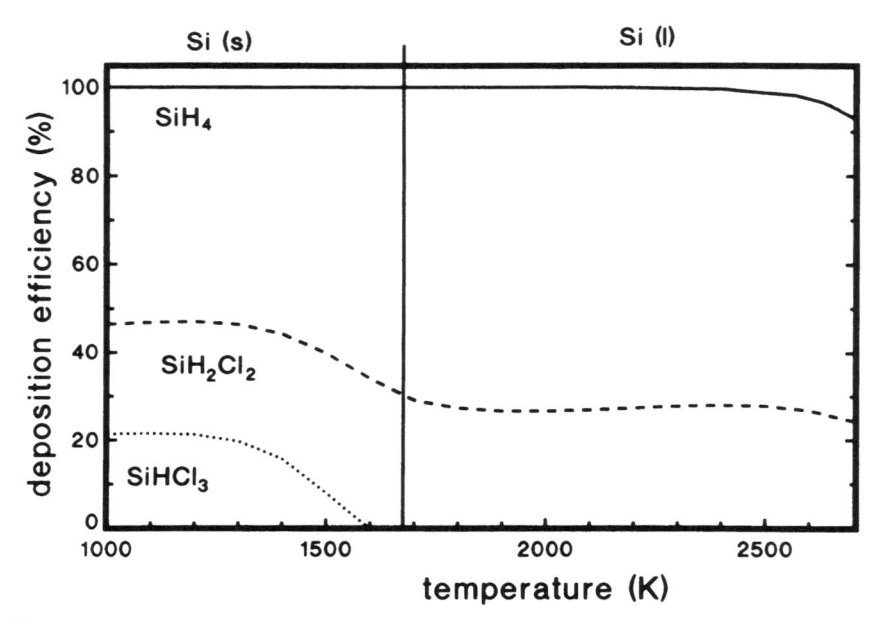

Fig. 2.3 Calculated thermodynamic deposition efficiency of solid and liquid from several silanes at atmospheric pressure. Silicon deposition does not occur at 10^5 Pa pure $SiCl_4$.

with the chlorinated silanes the deposition efficiency reduces with increasing chlorine content. It can be calculated that when using SiH_2Cl_2, silicon can be formed with an efficiency of about 46% which reduces at a temperature around 1400 K to 27%. Above a temperature of 2900 K the deposition efficiency of SiH_2Cl_2 reduces to zero. The calculations with SOLGASMIX reveal the maximum efficiency for the formation of silicon from $SiHCl_3$ to be about 20% at temperatures up to 1300 K and to decrease to zero when the temperature increased to 1500 K. The formation of silicon from pure $SiCl_4$ is thermodynamically not possible at atmospheric pressure. However, in these calculations the values of $Si(Cl_2)_n$, and other complex silicon–chlorine compounds, have not been included although they are known to be formed at moderate temperatures [26, 27]. Therefore, at moderate temperatures, the deposition efficiencies of silicon may be presumed to be lower. Hunt [29] has calculated the final silicon chlorine ratio, $(Si/Cl)_f$, in the gas phase for Si–H–Cl systems, if the system is in thermodynamic equilibrium. From this value the theoretical maximum deposition efficiencies of silicon, η, can be calculated, taking into account the initial silicon chlorine ratio, $(Si/Cl)_i$:

$$\eta = \left[1 - \frac{\left(\dfrac{Si}{Cl} \right)_f}{\left(\dfrac{Si}{Cl} \right)_i} \right] 100\%. \qquad (2.6)$$

Table 2.1 Thermodynamic deposition efficiencies of Si compared to powder yields in laser chemical vapour precipitation[a]

	η at < 1400 K	η at 1600 K	Powder yield from laser-CVP
SiH_2Cl_2	+46	+30	10
$SiHCl_3$	+19	−11	0
$SiCl_4 + 2H_2$	−8	−40	0
$SiCl_4 + 2N_2$	−8	−60	0

If $\eta < 0$ etching of silicon occurs instead of deposition
[a] Based on data of Hunt [32]

If this value is negative, only etching of silicon will occur. The deposition efficiencies, η, for the chlorinated silanes are shown in Table 2.1. These values are comparable with the present values. For $SiHCl_3$, without additional H_2, only etching will occur beyond 1600 K. The effect of the addition of H_2 is relatively small according to the calculations of Hunt. Only by adding 10 to 100 times as many hydrogen atoms as silicon atoms does the yield of silicon increase significantly.

Extensive thermodynamic calculations for the deposition of Si_3N_4 have been reported by Kingon *et al.* [8]. They calculated chemical vapour deposition (CVD) phase diagrams with $Si/(Si + N)$ ratios as a function of temperature up to values of 1800 K using several halogenated silanes as input gas at different pressures. Kruis *et al.* [30–32] have calculated CVD phase diagrams with silicon reactants and at the conditions used here to produce Si_3N_4. They reported CVD phase diagrams for SiH_4, SiH_2Cl_2, $SiHCl_3$ and $SiCl_4$ at atmospheric pressure and for temperatures up to 2600 K. In both studies SiH_4, reacting with NH_3, is practically completely transferred into condensed species. In the CVD phase diagrams a large region is shown with mixed Si and Si_3N_4 phases. Above 2033 K only Si is formed. On the other hand, the diagrams of the chlorinated silanes reveal that with increasing chlorine content the efficiencies of the condensed species as well as the regions with condensed phases other than Si_3N_4 are reduced. Therefore, minor silicon contamination is expected to occur with $SiCl_4$ as compared to SiH_4. Although the deposition efficiencies in both studies are similar, the deposition efficiencies in the CVD phase diagrams of $SiCl_4$ and SiH_2Cl_2 of Kingon *et al.* seem to behave irregularly, whereas the efficiencies reported by Kruis *et al.* are more continuous. Therefore, it seems that the figures from Kruis *et al.* are more plausible. Kingon *et al.* showed that addition of hydrogen may increase the deposition efficiencies of the condensed species, but it also increases the silicon contamination. Deposition efficiencies of the condensed phases were calculated based on a mixture of NH_3 and of one of the silanes SiH_4, SiH_2Cl_2, $SiHCl_3$ and $SiCl_4$ with an initial $Si/(Si + N)$ ratio as used in the present experiments. This section of CVD phase diagrams at atmospheric pressure

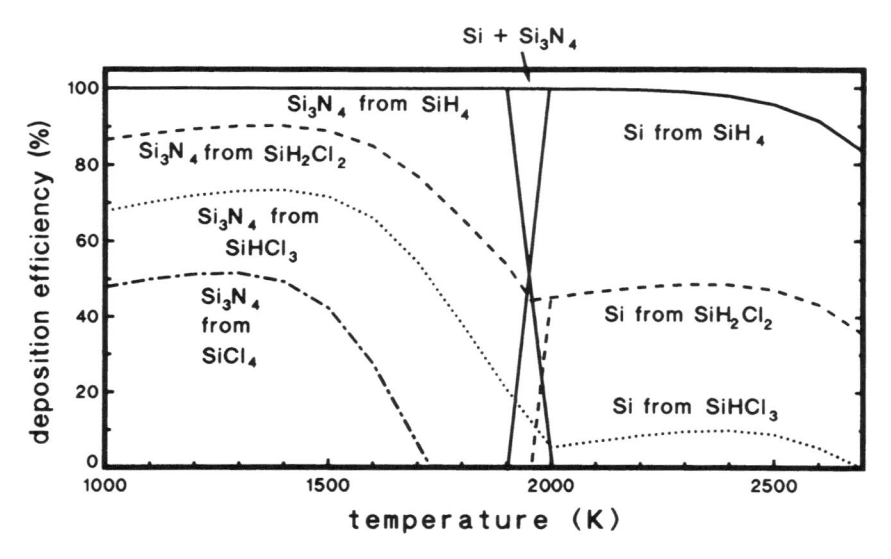

Fig. 2.4 Calculated thermodynamic deposition efficiency of Si_3N_4 and liquid silicon from several silanes with NH_3, $Si/(Si + N) = 1/3$, at atmospheric pressure.

with a $Si/(Si + N)$ ratio of 0.33 is shown in Fig. 2.4. While SiH_4 is virtually completely converted into either Si_3N_4 or liquid silicon at this $Si/(Si + N)$ ratio, the chlorinated silanes reveal, in addition to a decrease in Si_3N_4 deposition efficiency with increasing chlorine content of the reactant, also a strong reduction of silicon formation at higher temperatures. From the mixtures of $SiHCl_3$–NH_3 and $SiCl_4$–NH_3 with ratios 1:2, little or no deposition of silicon occurs thermodynamically at atmospheric pressure.

The temperature reached in a laser-CVP reactor is determined by the amount of energy absorbed from the laser radiation. With the use of a stable sensitizer, the amount of energy absorbed is thus especially independent of the reactants used. The laser heated process is, therefore, more an energy-regulated process than a temperature-regulated process. If the heat loss is neglected, the resulting temperature is determined by the heat capacity of the system and the heat necessary for the conversion of the reactants. Here the equilibrium composition was used as the resulting phase. Although the resulting practical mixture may deviate considerably from equilibrium conditions, the relation between input energy and equilibrium temperature provides information about the variation in the resulting temperature of flames containing different silanes. In Fig. 2.5 the energy required to reach equilibrium at a certain temperature from reactants at 298 K is shown. From this graph it is clear that, with increasing chlorine content of the reactant, the resulting temperature is lower. This is not only caused by an increase in heat capacity but especially by the higher enthalpy of formation of chlorine containing molecules. The formation enthalpy varies from $662.833 \, kJ \, mol^{-1}$ for $SiCl_4$ to

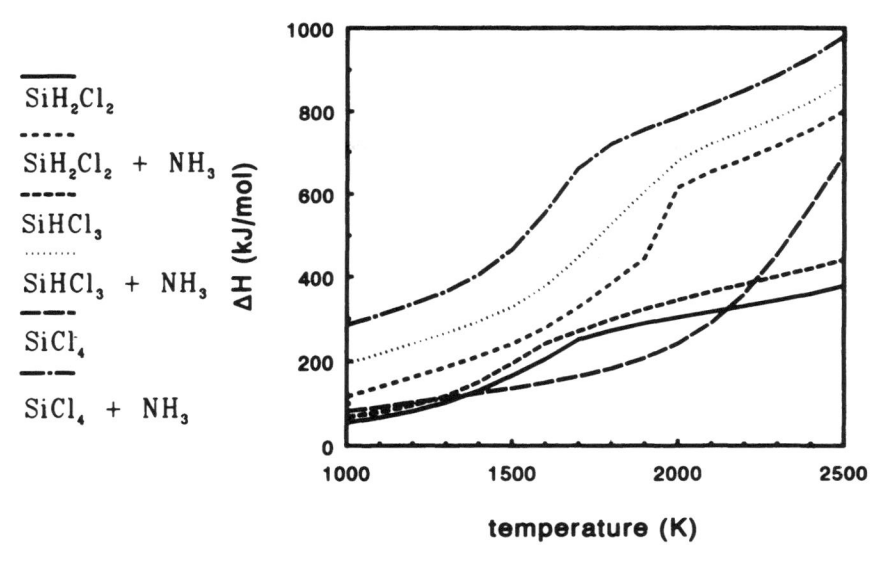

SiH₂Cl₂

SiH₂Cl₂ + NH₃

SiHCl₃

SiHCl₃ + NH₃

SiCl₄

SiCl₄ + NH₃

Fig. 2.5 The calculated enthalpy change, per mole chlorinated silane, from the reactant mixture at room temperature to thermodynamic equilibrium conditions at temperature.

$-34.309\,\mathrm{kJ\,mol^{-1}}$ for SiH_4 at ambient conditions, although the differences between the average bonding enthalpies are relatively small: $400\,\mathrm{kJ\,mol^{-1}}$ for the Si–Cl bond compared with $322\,\mathrm{kJ\,mol^{-1}}$ for the Si–H bond [23]. A relatively large amount of energy is released when molecular hydrogen is formed from the atomic species. Although the reaction of SiH_2Cl_2 with NH_3 is exothermic, the resulting temperature with the same amount of energy absorption is lower than is reached in the dissociation of pure SiH_2Cl_2. This can be explained by the additional energy required to heat NH_3. The reaction of $SiHCl_3$ and $SiCl_4$ with NH_3 is even endothermic, resulting in a decrease in the resulting temperature. The low position of the line of pure $SiCl_4$ and its distinct line shape are remarkable. This line shape can be explained by the high stability of pure $SiCl_4$ compared to the other two chlorinated silanes.

It is also interesting to calculate the effect of the addition of the sensitizers to a system with one of the chlorinated silanes and to a system with one of the chlorinated silanes and NH_3. The systems [chlorinated silane]–[sensitizer], 1:0.2, and [chlorinated silane]–NH_3–[sensitizer], 1:2:0.2 were calculated using different chlorinated silanes and sensitizers. According to thermodynamical calculations by SOLGASMIX, SF_6 is quite stable at atmospheric pressure and high temperatures up to 2000 K. However, when one of the chlorinated silanes or NH_3 is added to the system, SF_6 is completely dissociated. With the addition of a chlorinated silane, fluorine from SF_6 reacts, either with silicon to form SiF_4, or with hydrogen to form HF. Depending on the remaining fluorine, silicon and hydrogen concentrations it can be calculated that the sulphur reacts predominantly to SF_x (with x between 1 and 6), SiS, H_2S and S_2. By contrast, SiF_4 is thermodynamically much more stable. Negligible dissociation occurs, when SiF_4 is used instead of SF_6 with the chlorinated silanes. At temperatures above 1600 K dissociation of SiF_4 becomes noticeable. Noticeably, the addition of NH_3 to SiF_4 results in an increased dissociation of the sensitizer at temperatures beyond 1600 K.

2.3.3   Visible spectroscopy

When a reactant mixture was injected into the reactor and irradiated by the CO_2-laser, luminescence from both particulate and gaseous products was observed as a flame [17]. The colour and light intensity of the flame varied strongly with the reactant and the sensitizer used. While, when using SiH_2Cl_2 as silicon reactant, a bright flame was obtained both with and without NH_3, the other two chlorinated silanes, $SiHCl_3$ and $SiCl_4$, resulted in a weak flame. Therefore, the visible spectra of these flames were explored to gain information on the nature of the luminescence.

To check the stability of the sensitizers against H_2 and NH_3, visible spectra of the mixtures SF_6–H_2, SF_6–NH_3, SiF_4–H_2 and SiF_4–NH_3 were recorded. The observed bands are summarized in Table 2.2. while excitation of the pure sensitizers only resulted in very weak luminescence, irradiating both sensitizers in the presence of H_2 resulted in much stronger luminescence. Especially, upon irradiating SF_6 in the presence of a low concentration of H_2, $H_2/SF_6 = 4$, a very strong blue emission was observed. The visible spectra of this flame contained emissions from 300 to 500 nm, degrading to the infrared. The bands compared well with the bands from the transition $B^3\Sigma_u^- - X^3\Sigma_g^-$ of S_2 [33]. This luminescence increases with increasing H_2 pressure, indicating a strong dissociation of SF_6 in the presence of H_2, as has been observed by other authors [18, 19]. Upon mixing SF_6 with NH_3 the visible spectrum of the flame contained in addition to this S_2 band an extensive number of emission bands

Table 2.2 Spectral analysis of several sensitizer gas mixtures excited by CO_2-laser radiation

Sensitizer	Other gas	Observed bands λ (nm)	Possible species
SF_6	—	little emission	
SiF_4	—	band 500–750	$SiF_3^+ + F^-$?
SF_6	H_2	400–600	S_2
SiF_4	H_2	band 500–750	$SiF_3^+ + F^-$?
		peak 440	SiF˙
SF_6	NH_3	400–500	S_2
		bands 500–800	NH_2˙
SiF_4	NH_3	440	SiF˙
		bands 500–800	NH_2˙

from 500 to 800 nm. This latter series of bands can be attributed to NH_2^* radicals. A similar result was observed, when SiF_4 was studied. While with pure SiF_4 a very weak broad emission feature was observed betwen 500 and 750 nm, addition of H_2 resulted in a strong emission peak centred at 437 nm. The origin of the broad emission feature is not clear, but Yahav and Haas [34] suggested an intermediate dissociation of SiF_4 into $SiF_3^+ + F^-$ by an inverse electronic relaxation mechanism, that is internal conversion of vibrational excitation to electronic excitation. From this electronically excited state a radiative transfer can occur to a state with an intermediately separated radical pair $SiF_3^· + F^·$. The peak in these spectra can be attributed to the $A^2\Sigma^+ - X^2\Pi$ transition of the $SiF^·$ radical [33]. By adding NH_3 instead of H_2 to SiF_4, an emission from $SiF^·$ and $NH_2^·$ radicals was observed. Therefore, both sensitizers are not stable with respect to H_2 and NH_3, when they are irradiated by CO_2-laser light.

The visible spectra were recorded from laser flames containing one of the chlorinated silanes SiH_2Cl_2, $SiHCl_3$ and $SiCl_4$ with one of the sensitizers SF_6 and SiF_4. Also the spectra from the same mixtures with NH_3 were recorded. One may expect to observe black body radiation from condensed phases, i.e. solid Si_3N_4 and solid and liquid silicon, in addition to luminescence of intermediates. The luminescence bands were compared to data in Pearse and Gaydon [33] and Suchard et al. [35, 36]. The resulting spectra, with SiH_2Cl_2 as chlorinated silane, are shown with fitted black body spectra in Fig. 2.1. The flow rates used for SiH_2Cl_2, the sensitizer and NH_3 were 53, 10 and 110 sccm, respectively. The temperatures of the fitted black body curves with the observed emission bands are summarized in Tables 2.3 and 2.4. While excitation of the systems SiH_2Cl_2 and $SiH_2Cl_2-NH_3$ with SiF_4 resulted in smooth black body spectra, excitation of the same systems with SF_6 leads to additional bands. In both the systems SiH_2Cl_2 and $SiH_2Cl_2-NH_3$, with SF_6 as sensitizer,

Table 2.3 Spectral analysis of several gas mixtures excited by CO_2-laser radiation

Si reactant	Sensitizer	Temperature of Si particles (K)	Observed bands λ (nm)	Possible species
SiH_2Cl_2	—	1700	—	
SiH_2Cl_2	SF_6	2180	440	$SiF^·$
			300–550	S_2
SiH_2Cl_2	SiF_4	1800	—	
$SiHCl_3$	SF_6		400–550	S_2
$SiHCl_3$	SiF_4		440	$SiF^·$
			680–750	a
$SiCl_4$	SF_6		450–650	$SiCl^·$
			680–760	a
$SiCl_4$	SiF_4		500–700	b
			680–750	a

[a] and [b] represents different unidentified bands

Table 2.4 Spectral analysis of several gas mixtures with NH_3 excited by CO_2-laser radiation

Si reactant	Sensitizer	Temperature of particles (K)	Observed bands λ (nm)	Possible species
SiH_2Cl_2	—	2630	—	
SiH_2Cl_2	SF_6	2620	440	$SiF^{\cdot}$
			300–550	S_2
SiH_2Cl_2	SiF_4	2300	—	
$SiHCl_3$	SF_6		440	$SiF^{\cdot}$
			680–750	a
$SiHCl_3$	SiF_4	2850[c]	—	
$SiCl_4$	SF_6		500–700	b
$SiCl_4$	SiF_4		500–800	$NH_2^{\cdot}$

[a] and [b] represent different unidentified bands
[c] denotes a poor black body fit

an emission peak centred at 437 nm was observed. This emission can be ascribed to the $A^2\Sigma^+-X^2\Pi$ transition of the $SiF^{\cdot}$ radical [33]. At lower wavelengths a difference between the fitted black body spectrum and the recorded spectrum is observed. This indicates the existence of a gas emission band. This band, which decreases in intensity from 300 nm towards 550 nm, was also observed in flames with SF_6 and H_2. Therefore, this band may be attributed to S_2. Whereas, in a flame of SiH_2Cl_2 with SF_6 at low flow rates (below 6.7 sccm, a thermal flame colour (yellow white) was always observed, at a high SF_6 flux ($\pm$ 17 sccm) a blue luminescence could be seen beyond the reaction flame core. Similar luminescence has been observed [18, 19, 22] and ascribed to luminescence of sulphur particles [19] formed during decomposition of laser excited SF_6. These references corroborate with the present spectroscopical analysis.

The flames with chlorinated silanes other than SiH_2Cl_2 were much less stable. With the use of $SiCl_4$ the flow fluctuated which was due to bubbling the carrier gas through the $SiCl_4$ liquid in the gas supply system. Additionally, the amount of luminescence was considerably lower. Consequently, the resolution of the recorded spectra from these mixtures attained is considerably lower. To ascertain that the observed bands were not caused by noise, longer integration times were used, and the spectra were recorded at least three times to verify the emission peaks. Because of the reduced emission only the spectrum from 400 up to 80 nm was recorded. The observed emission bands are summarized in Tables 2.3 and 2.4. The flow rates used for $SiHCl_3$ or $SiCl_4$, the sensitizer and NH_3 were 30, 10 and 60 sccm, respectively. Of all the spectra produced with $SiHCl_3$ or $SiCl_4$ only one could be compared with a black body spectrum, i.e. the spectrum from a mixture of $SiHCl_3-NH_3-SiF_4$, and even this fit was far from perfect. In adition to the emission bands from $SiF^{\cdot}$ and S_2 mentioned above, two unknown emission features were recorded in the spectra with $SiHCl_3$, i.e. a small emission feature between 680 and 770 nm, and a feature emission between 500 and 700 nm, degrading to the blue. It seemed that these bands were also present in the spectrum from the mixture of $SiHCl_3$–NH_3–SiF_4. The presence of these bands prevented an optimal fit of a black body curve. However, the spectra recorded from flames containing $SiHCl_3$ indicate that both sensitizers are not completely stable. Only in the system $SiHCl_3$–NH_3–SiF_4 could black body emission from a condensed phase be observed.

None of the emission bands that can be attributed to dissociation products of one of the sensitizers were observed in the spectra of flames containing $SiCl_4$. Neither did the recorded spectra contain curves that could be matched with black body radiation. From the flame of $SiCl_4$ and SF_6 a large emission feature was observed extending from 450 to 650 nm and with a maximum around 520 nm. This emission band compares well with the literature data on the transition $A^2\Sigma^+$–$X^2\Pi$ of $SiCl^\cdot$ [33] which indicates that $SiCl_4$ is partially dissociated under these conditions. From the flame containing $SiCl_4$–NH_3–SiF_4 an emission feature, that can clearly be attributed to $NH_2^\cdot$ was observed. In the other flames with $SiCl_4$, only the previously mentioned unknown emission features at 500–800 μm and 680–750 μm were observed. Hence there is no spectroscopic evidence that $SiCl_4$, either with or without NH_3, reacts to form a condensed phase. However, there is also no indication that either of the sensitizers dissociates in the presence of $SiCl_4$.

2.3.4   Synthesis of silicon

Excitation of $SiHCl_3$ and $SiCl_4$ with either SF_6 or SiF_4 did not result in the formation of any condensed phase. In addition, neither silicon nor the solid by-product $(SiCl_2)_n$ was collected in the precipitator. Therefore, SiH_2Cl_2 was separately excited in the presence of SF_6, SiF_4 and N_2 to evaluate differences in the condensed products formed. While with SiF_4 the laser line P42 of the 9.4 μm band was used, for the other two mixtures with N_2 and SF_6 the laser line P20 of the 10.4 μm band was used to excite SiH_2Cl_2 directly and in the presence of the sensitizer mainly indirectly, respectively. The flow rate of SiH_2Cl_2 was set at 53 sccm and that of the additional gas was set at 10 sccm. The temperatures measured in the experiments with SiF_4 and with N_2 were relatively low, i.e. 1670 and 1630 K, respectively, whereas with SF_6 a considerably higher flame temperature of 1830 K was measured by two-colour pyrometry. From all three experiments a brownish solid product was collected in the electrostatic precipitator. The powder was obtained from the precipitator with yields of about 10%, based on the silicon consumption. The product synthesized with SF_6 was more yellow and revealed the very odorous smell of H_2S. Most of the particles from the experiments with SiF_4 and N_2 were crystalline to both X-ray diffraction and to electron diffraction on the transmission electron microscope. The diffraction pattern compared well with the pattern of silicon. On the other hand, the product from the SiH_2Cl_2–SF_6 mixture did not reveal any crystallinity. Infrared transmission of the particles revealed, next to the Si–O stretching band at a wavenumber of $1090\,cm^{-1}$ an absorption at $580\,cm^{-1}$. When SiF_4 was utilized as sensitizer, the product revealed an especially, strong absorption near to this wavenumber. This absorption can be attributed to a Si–Cl band [10]. The product formed from the SiH_2Cl_2–SF_6 mixture contained a considerable amount of sulphur (15wt%), indicating that a large amount of the decomposition product of the sensitizer was incorporated into the product. The amount of chlorine in the same product was considerably lower, i.e. 7wt%. With the use of the other sensitizer, SiF_4, the large amount, 29wt%, of chlorine was incorporated into the powder. This confirms the results of IR-spectroscopy. This large amount of chlorine might indicate the formation of $(SiCl_2)_n$. The electron micrograph of Fig. 2.6 shows two classes of particles from the systems SiH_2Cl_2–SiF_4 and SiH_2Cl_2–N_2: non-crystalline, strongly agglomerated particles with diameters around 35 nm, and crystalline particles with diameters between 60 and 100 nm. In contrast, the powder from the system SiH_2Cl_2–SF_6 contained only small particles around 50 nm. In these powders sulphur contamination was observed as a separate phase.

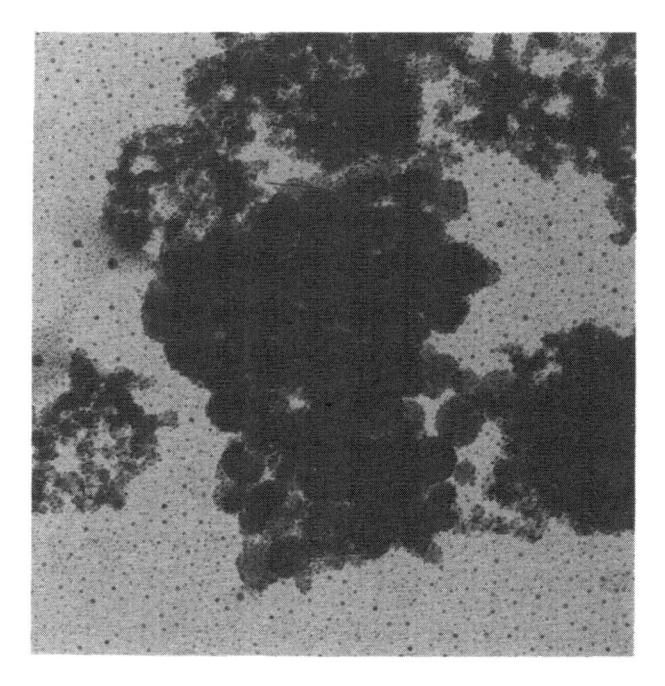

Fig. 2.6 Transmission electron micrograph of crystalline silicon particles from the dissociation of SiH_2Cl_2 sensitized by SiF_4.

2.3.5 Synthesis of Si_3N_4

The characteristics of the Si_3N_4 powders formed from the different chlorinated silanes and sensitizers are shown in Table 2.5. When using SiF_4 the laser line P42 of the 9.4 μm band was used to heat the gases, while for the other mixtures with SF_6 the laser line P20 of the 10.4 μm band was used.

Si_3N_4 powders were prepared from SiH_2Cl_2 and NH_3 by sensitized reactions with both SF_6 and SiF_4. The flow rates of the sensitizer, SiH_2Cl_2 and NH_3 were set at 10, 53 and 110 sccm, respectively. A light-brown powder was collected in the precipitator. Despite the fact that a considerable part of the synthesized powders had settled together with NH_4Cl in the transit pipe between the reactor and the precipitator, the yield of powders recovered from the precipitator was still around 25% in both experiments. Hence, yields of more than 25% can be achieved. Transmission electron microscopy revealed spherical particles with diameters, d, around 70 nm. In both experiments fibres were also formed such as were observed from pure mixtures of SiH_2Cl_2 and NH_3 [6]. With infrared spectroscopy small amounts of imides (2900 cm^{-1} and 1200 cm^{-1}) and NH_4Cl (1400 cm^{-1}) were detected along with Si_3N_4 (950 cm^{-1}). X-ray diffraction revealed the silicon nitride powders to be amorphous. By BET surface areas, A, of the order of $100 \, m^2 g^{-1}$ were measured. Less than 0.5% sulphur was detected, using X-ray fluorescence, in the Si_3N_4 powder from the gas mixture with SF_6. However, in preliminary experiments, the use of much higher concentrations of SF_6 resulted in crystalline sulphur particles as determined by X-ray diffraction.

The same experiments were performed utilizing $SiHCl_3$ and $SiCl_4$ as silicon source for the production of Si_3N_4. Whereas SiH_2Cl_2 can absorb some lines of the CO_2-laser, $SiHCl_3$ and $SiCl_4$ can only be heated by energy transfer from the excited sensitizer. In the experiments with $SiCl_4$, some of the powder spread through the reactor instead of leaving through the exhaust funnel. This powder is likely to be a low temperature imide product similar to the powder formed without using the laser. Because the low-temperature product is not

Table 2.5 Process conditions and powder characteristics of formed Si_3N_4

Si reactant	Sensitizer	Flow rate NH_3 (sccm)	Area from BET ($m^2 g^{-1}$)	d from TEM (nm)	IR absorption NH-band	N (wt%)
SiH_2Cl_2	SiF_4	110	91	26	medium	24
$SiHCl_3$	SF_6	60	—	30	medium	21
$SiHCl_3$	SiF_4	60	145	40	weak	32
$SiCl_4$	SF_6	60	—	15,40	strong	—
$SiCl_4$	SiF_4	60	105	30,110	strong	17

Flow rate sensitizer: 10 sccm
Flow rate SiH_2Cl_2: 53 sccm
Flow rate $SiHCl_3$ and $SiCl_4$: 30 sccm

formed in a strongly upward directed gas stream, it spreads in the reactor more easily. In particular, with $SiCl_4$ the amount of transmitted laser radiation was also very low compared to previous experiments. The transmission was only 0.59 compared to the typical transmission, which is in the order of 0.85. Probably the laser radiation was scattered by the imide powder which was spread in the reactor. In the experiments with $SiHCl_3$ and $SiCl_4$ as silicon source, lower yields of the order of 10% were obtained from the precipitator. The analysis results showed little difference between all these powders with variation of the silicon source or of the sensitizer. The powders collected were light tan coloured and the particles were quite small, i.e. 15 to 40 nm, when observed in the TEM (Figs 2.7 and 2.8). Micrographs of these powders revealed only spherical particles in contrast to micrographs of the powders from SiH_2Cl_2. TEM pictures showed that the powders from $SiCl_4$ also contained some larger particles with a diameter of about 110 nm. Experiments in which $SiCl_4$ and NH_3 reacted without using a laser beam revealed only those large particles. This suggests that these larger particles consist of silicon diimide due to low-temperature reactions. IR-spectroscopy also confirmed the presence of a considerable amount of imides ($2900 \, cm^{-1}$ and $1200 \, cm^{-1}$) in

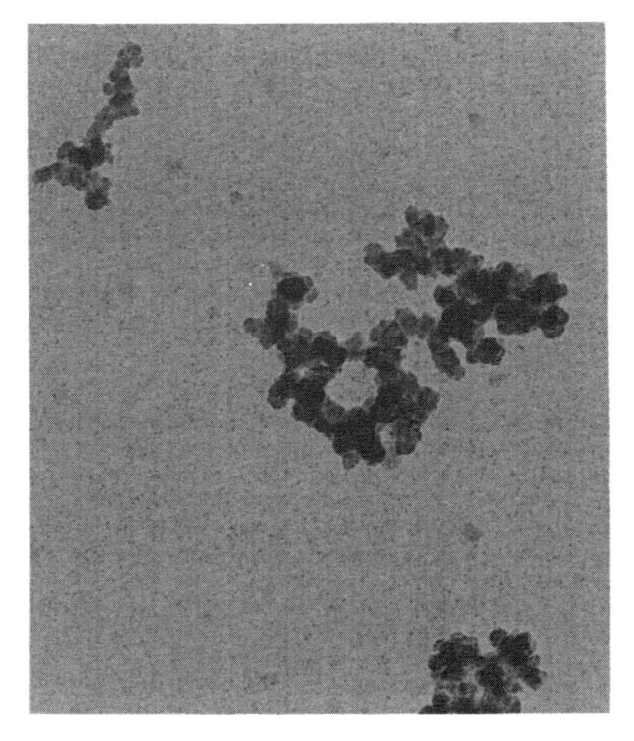

Fig. 2.7 Transmission electron micrograph of Si_3N_4 particles from a mixture of SiF_4, $SiHCl_3$ and NH_3.

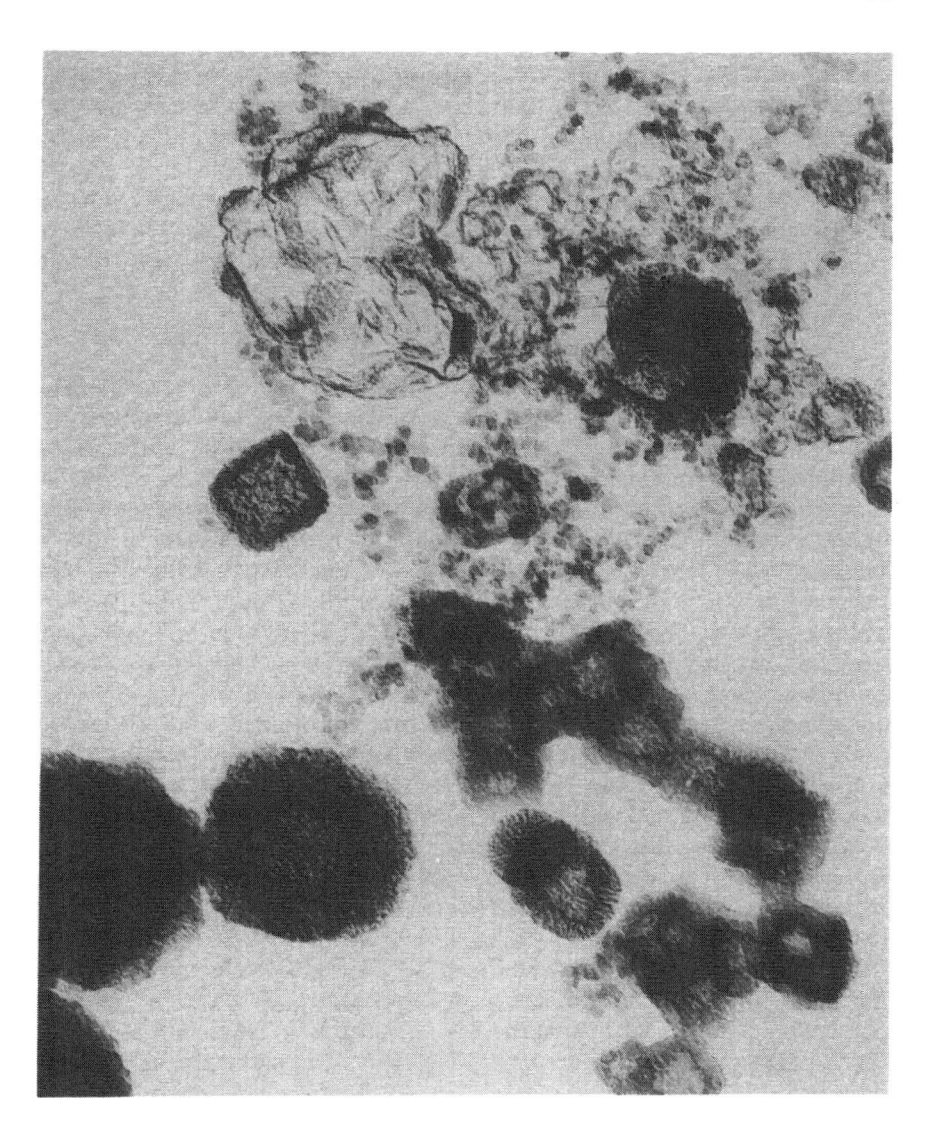

Fig. 2.8 Transmission electron micrograph of particles precipitated from a mixture of SF_6, $SiCl_4$ and NH_3. The small particles are Si_3N_4, the larger particles are silicon imides and the glassy particle is NH_4Cl.

addition to the silicon nitride ($950\,cm^{-1}$), in the powders synthesized from $SiCl_4$, compared to a moderate amount in the powders prepared from $SiHCl_3$. Typically, the powders collected from $SiCl_4$ contained about 3wt% chlorine as measured by microtitration, whereas the powders prepared from $SiHCl_3$ contained between 0.3 and 1.5wt% chlorine. The use of SF_6 as a sensitizer resulted in a contamination of about 2wt% sulphur as measured by ICP in the reaction products from both $SiHCl_3$ and $SiCl_4$. Although energy dispersive X-ray spectroscopy, EDX, on the TEM revealed practically zero fluorine contamination, a small amount of $(NH_4)_2SiF_6$ was observed in the powders synthesized from $SiCl_4$ and NH_3 with SF_6 by X-ray diffraction. Therefore, also in this experiment, SF_6 was not completely inert. Crystalline Si_3N_4 was not detected with X-ray diffraction in any of the powders synthesized here.

2.4 DISCUSSION

Although the powders produced from SiH_2Cl_2 with either N_2 or SiF_4 contained mainly crystalline silicon, a considerable chlorine contamination was detected in the particles produced from the mixture SiH_2Cl_2–SiF_4. In the powders from SF_6 a considerably lower chlorine contamination was observed. In preliminary experiments a much lower laser energy of tens of watts was used to dissociate SiH_2Cl_2 compared to 130 W in the above mentioned synthesis. This resulted in a yellow-brown, fluffy powder which reacted violently with either air or water in the presence of a strong luminescence. These powders formed with low laser power, and consequently produced at moderate temperatures, are likely to contain unstable compounds such as silicon chlorine compounds. Apparently, at those moderate temperatures SiH_2Cl_2 does not dissociate completely. SiH_2Cl_2 has also been dissociated using CO_2-laser radiation by other researchers [37]. By exciting SiH_2Cl_2 at low pressures, and with short CO_2-laser pulses, Sausa and Ronn [37] observed emission of $SiCl_2$ ($^1B_1 \rightarrow {}^1A_1$) around a wavelength of 340 nm as well as particulate formation from HeNe-laser scattering. Although these authors did not analyse the particulates formed, they concluded from the emission spectra that SiH_2Cl_2 dissociated into $SiCl_2$ and that these dissociation products polymerized to form particulates comprising $(SiCl_2)_n$. Because they used low pressures (2 Pa $< P <$ 140 Pa), and a short pulse length (300 ns), they concluded that their dissociation reaction was primarily monomolecular. Here the collision frequency is $6 \times 10^9\,s^{-1}$. Thus, a SiH_2Cl_2 molecule will collide about 8×10^6 times in the laser beam. Therefore, in the present experiments, the reaction can be expected to be far from monomolecular. This may explain the difference in product character between powders obtained in the present experiments at high laser power density, and the powders obtained by Sausa and Ronn [37].

Both visible spectroscopy and the synthesis results reveal that there is a large difference between the use of SiH_2Cl_2 and the use of either $SiHCl_3$ or $SiCl_4$ as silicon source. As expected from the thermodynamical calculations, and confirmed by visible spectroscopy as well as by the absence of the condensed products in the precipitator, neither $SiHCl_3$ nor $SiCl_4$ were converted into solid silicon. Because silicon cannot be formed from these reactants the possible reaction mechanisms for the formation of Si_3N_4 are reduced. Larger Si_3N_4 particles can be formed from SiH_2Cl_2 by the initial formation and growth of liquid silicon droplets [17] which are subsequently nitrided by NH_3. Since the experiments do not reveal this initial formation of silicon droplets from $SiHCl_3$ and $SiCl_4$, this particular mechanism for the formation of Si_3N_4 can be disregarded. The Si_3N_4 particles formed from either $SiHCl_3$ or $SiCl_4$ reveal similar properties as those from SiH_2Cl_2 with the distinction that in the latter case fibres were also formed. This fibre formation is described in more detail elsewhere [6].

The amount of chlorine contamination is similarly low for the different Si_3N_4 powders. The main distinction between the different silicon reactants is the amount of imide contamination. With the use of $SiCl_4$ this imide contamination was considerably increased. Thermodynamically, two causes may account for this increase. First, from silanes with an increasing content of chlorine a lower yield of Si_3N_4 is obtained, as is shown in Fig. 2.4. Second, with the use of $SiCl_4$, the reaction temperature decreases with the same laser power, as is shown in Fig. 2.5. This lower temperature may result in a decreased conversion of silicon imides into silicon nitride. This may result in a relatively large amount of imide compared to nitride.

Although the thermodynamic calculations reveal large differences in the stability between the two sensitizers in the presence of H_2 and NH_3, visible spectroscopy reveals that both sensitizers seem to be dissociated to some extent in the presence of the hydrogen rich reactants. Whereas the visible spectra of SF_6 with chlorinated silanes show emissions from dissociation products, SiF_4 demonstrates little or no emission from dissociation products in the same environment. According to the thermodynamic calculations fluorine from dissociation of both sensitizers can react with the reactants to form HF and SiF_x (with x between 1 and 4). This latter product may lower the production yield of either silicon or Si_3N_4. While the dissociation of SiF_4 may result in additional silicon in the product, the dissociation of SF_6 will result in sulphur contamination in the powders produced. The dissociation of the sensitizer SiF_4 is, therefore, much less harmful for the production of Si_3N_4 and silicon than the dissociation of SF_6.

2.5 CONCLUSIONS

The sensitized formation of both silicon and Si_3N_4 powders using SiH_2Cl_2 as silicon source is shown to be feasible. While the dissociation of both $SiHCl_3$ and $SiCl_4$ to silicon powders has not been successful, the sensitized heating of these chlorinated silanes with NH_3 resulted in the formation of Si_3N_4 powders with comparable yield. Of the sensitizers used SF_6 shows a considerably stronger absorption. Both sensitizers reveal instability when irradiated with hydrogen-rich reactants, e.g. NH_3, SiH_2Cl_2, and $SiHCl_3$. However, by thermodynamic calculations, spectroscopic measurements and product analysis SiF_4 is shown to cause considerably less contamination than SF_6 in the powders formed.

REFERENCES

1. Barringer, E.A. and Bowen, H. K. (1984) Ceramic powder processing. *Ceram. Eng. Sci. Proc.*, **5**, 285–97.
2. Jacquemijns, E.J., van der Put, P.J. and Schoonman, J. (1988) Vapour phase synthesis of ultrafine silicon nitride powders. *High Temp. High Press.* **20**, 31–4.
3. Cannon, W.R., Danforth, S.C., Flint, J.H., Haggerty, J.S. and Marra, R.A. (1982) Sinterable ceramic powders from laser-driven reactions: I, process description and modeling. *J. Am. Ceram. Soc.*, **65**, 324–30.
4. Cannon, W.R., Danforth, S.C., Haggerty, J.S. and Marra R.A. (1982) Sinterable ceramic powders from laser-driven reactions: II, powder characteristics and process variables. *J. Am. Ceram. Soc.*, **65**, 330–35.
5. Sawano, K., Haggerty, J.S. and Bowen, H.K. (1987) Formation of SiC powder from laser heated vapor phase reactions. *Yogyo Kyokai Shi*, **95**, 64–9.
6. Lihrmann, J.M., Haggerty, J.S., Luce, M., Croix, O. and Cauchetier, M. (1987) Potentiel des céramiques thermomécaniques élaborées à partir de poudres laser. (Potential of thermomechanical ceramics made from laser powders). *Matériaux Méchanique Électricité*, **422**, 32–6.
7. Symons, W. and Danforth, S.C. (1987) Synthesis and characterization of laser synthesized silicon nitride powders, in *Advances in Ceramics*, vol. 21 (eds. G.L. Messing, K.S. Mazdiyasni, J.W. McCauley and R.A. Haber), American Ceramic Society, Westerville Ohio, USA, pp. 249–56.
8. Kingon, A.I., Lutz, L.J. and Davis, R.F. (1983) Thermodynamic calculations for the chemical vapor deposition of silicon nitride. *J. Am. Ceram. Soc.*, **66**, 551–8.
9. Mazdiyasni, K.S. and Cooke, C.M. (1973) Synthesis, characterization, and consolidation of Si_3N_4 obtained from ammonolysis of $SiCl_4$. *J. Am. Ceram. Soc.*, **56**, 628–33.
10. Billy, M. (1959) Préparation et définition du nitrure de silicium. (Preparation and characterization of silicon nitride). *Ann. Chim. (Paris)*, **4**, 795–851.
11. Billy, M., Brossard, M., Desmaison, J., Giraud, D. and Goursat P. (1975) Synthesis of Si and Ge nitrides and Si oxynitride by ammonolysis of chlorides—comment on 'Synthesis, characterization, and consolidation of Si_3N_4 obtained from ammonolysis of $SiCl_4$'. *J. Am. Ceram. Soc.*, **57**, 254–5.
12. Calvert, J.G. and Pitts, Jr., J.N. (1966) *Photochemistry*. John-Wiley, New York, p. 19.
13. Shaub, W.M. and Bauer, S.H. (1975) Laser-powered homogeneous pyrolysis. *Int. J. Chem. Kinet.*, **7**, 509–29.
14. Steinfeld, J.I. (1981) Laser-induced chemical reactions: survey of the literature, 1965–1979, in *Laser-Induced Chemical Processes* (ed J.I. Steinfeld), Plenum Press, New York, pp. 243–67.
15. Brunet, H. (1970) Saturation infrared absorption in SF_6. *IEEE J. Quant. Electron.*, **6**, 678–84.
16. Nowak, A.V. and Lyman, J.L. (1975) The temperature-dependent absorption spectrum of the v_3 band of SF_6 at 10.6 μm. *J. Quant. Spectrosc. Radiat. Transfer*, **15**, 945–61.
17. Bauer, R.A. (1991) Laser chemical vapor precipitation of ceramic powders. PhD Thesis, Delft.
18. Ronn, A.M. (1976) Particulate formation by infrared laser dielectric breakdown. *Chem. Phys. Lett.*, **42**, 202–4.
19. Lin, S.T. and Ronn, A.M. (1978) Laser induced sulfur particulate formation. *Chem. Phys. Lett.*, **56**, 414–18.
20. Bauer, S.H. and Haberman, J.A. (1978) A laser augmented reaction:

$SF_6 + SiH_4 \rightarrow S_2^* + SiF_4 + HF + H_2\dot{}$, retention of isotopic selectivity during detonation. *IEEE J. Quant. Electron.*, **14**, 233–7.

21. Steinfeld, J.I., Burak, I., Sutton, D.J. and Nowak, A.V. (1970) Infrared double resonance in sulfur hexafluoride. *J. Chem. Phys.*, **52**, 5421–34.

22. Olszyna, K.J., Grunwald, E., Keehn, P.M. and Anderson, S.P. (1977) Megawatt infrared laser chemistry. II. use of SiF_4 as an inert sensitizer. *Tetrah. Lett.*, **19**, 1609–12.

23. Chase M.W. (ed) (1985) *JANAF Thermochemical Tables*, 3rd edn., National Bureau of Standards, Washington, DC.

24. Eriksson, G. (1975) Thermodynamic studies of high-temperature equilibriums. XII. SOLGASMIX, a computer program for calculations of equilibrium compositions in multiphase systems. *Chem. Scr.*, **8**, 100–103.

25. White, W.B., Johnson, W.M. and Dantzig, G.B. (1958) Chemical equilibrium in complex mixtures. *J. Chem. Phys.*, **28**, 751–5.

26. Schmeisser, M. and Voss P. (1964) Über das Siliciumdichlorid $[SiCl_2]_x$ (About silicon dichloride $[SiCl_2]_x$. *Z. Anorg. Allgem. Chem.*, **334**, 50–56.

27. Schenk, P.W. and Bloching, H. (1964) Darstellung und Eigenschaften des Siliciumdichlorids $(SiCl_2)_x$ (Preparation and properties of silicon dichloride $(SiCl_2)_x$). *Z. Anorg. Allgem. Chem.*, **334**, 57–65.

28. Puxbaum, H. and Vendl. A. (1977) Zur Relativkonduktometrischen Bestimmung von Stickstof in einigen Siliciumnitridhaltigen Werkstoffen (For relative conductrometric determination of nitrogen in several silicon nitride containing components). *Fresenius Z. Anal. Chem.*, **287**, 134–7.

29. Hunt, L.P. (1988) Thermodynamic equilibria in the Si–H–Cl and Si–H–Br systems. *J. Electrochem. Soc.*, **135**, 206–9.

30. Kruis, F.E., Scarlett, B.A., Bauer, R.A. and Schoonman, J. (1990) Homogeneous nucleation of silicon in an aerosol reactor using laser-heating. *Aerosols, Science, Industry, Health and Environment*. Proceedings of the Third International Aerosol Conference (eds S. Masuda and K. Takahashi), September 24–27, 1990, Kyoto, Japan. Pergamon Press, Oxford.

31. Bauer, R.A., Kruis, F.E., van der Put, P., Scarlett, B. and Schoonman, J. (1990) Chemical vapour precipitation of silicon nitride powders in a laser reactor. *KONA*, **8**, 145–54.

32. Kruis, F.E., Scarlett, B.A., Bauer, R.A. and Schoonman, J. (1992) Thermodynamic calculations on the chemical vapor deposition of silicon nitride and silicon from silane and chlorinated silanes. *J. Am. Ceram. Soc.*, **75**(3), 619–628.

33. Pearse, R.W.B. and Gaydon, A.G. (1976) *The identification of molecular spectra*, 4th edn, Chapman and Hall, London, UK.

34. Yahav, G. and Haas, Y. (1981) Silicon tetrafluoride visible luminescence induced by high-power CO_2 laser irradiation. *Chem. Phys. Lett.*, **83**, 493–497.

35. Suchard, S.N. (ed) (1975) Heteronuclear diatomic molecules, in *Spectroscopic Data*, Vol. 1, Plenum, New York.

36. Suchard, S.N., and Melzer, J.E. (ed) (1976) Homonuclear Diatomic Molecules in *Spectroscopic Data*, Vol. 2, Plenum, New York.

37. Sausa, R.C. and Ronn, A.M. (1985) Infrared multiple photon dissociation of dichlorosilane: the production of electronically excited $SiCl_2$. *Chem. Phys.*, **96**, 183–9.

CHAPTER 3

Particle interactions in suspensions

R.G. Horn

3.1 INTRODUCTION

The strength of ceramic materials is limited by the presence of defects or flaws. Such flaws commonly arise from **agglomerates** of particles present in the early stages of processing, which lead to packing inhomogeneities when the powders are compacted, and voids after firing. One effective method of minimizing the number and size of agglomerates, and also of mixing powders homogeneously, is to disperse particles by suspending them in a liquid. This is the so-called colloidal route to ceramic processing [1].

Typical steps involved in colloidal processing are preparation and dispersion of powders, compaction to a high volume fraction of solids by removing most of the suspending liquid, forming, further drying or burning out of remaining liquid and other processing ingredients, and firing. The primary focus of this chapter is on the dispersion of particles, but we will try to keep in mind the subsequent steps, in which factors such as rheology and the amount of liquid remaining in the suspension are very important considerations.

Solid particles generally have a higher density than the liquids that might be used to suspend them, so they are inclined to sink. Brownian motion, however, will keep particles suspended if they are sufficiently small. In that case we say the suspension is **stable**. To understand the stability of suspensions we also need to consider the forces that act between particles, which can be either attractive or repulsive depending on a variety of conditions to be discussed below. In simple terms, if the forces are repulsive then particles remain separated and small particles can stay suspended (dispersed) for long periods of time. On the other hand, attractive forces between particles cause small particles to stick together and form agglomerates,which may become large enough and heavy enough to fall out of suspension and form a sediment.

Ceramic Processing. Edited by R.A. Terpstra, P.P.A.C. Pex and A.H. de Vries.
Published in 1995 by Chapman & Hall, London. ISBN 0 412 59830 2

The behaviour of particle dispersions is therefore determined largely by the forces acting between the particles. For reasons to be discussed in the next section, these are often called **surface forces**. Apart from the basic **stability** of a suspension, surface forces also strongly affect its **rheology**. Clearly, both of these are vital factors in the colloidal processing of ceramics. A detailed understanding of the possible structures and phase behaviour of a particle suspension requires a full statistical mechanics treatment which includes consideration of interparticle forces, and a description of the rheology requires non-equilibrium statistical mechanics. We will not attempt to do the statistical mechanics here, but we will endeavour to describe the various forces that may be present. One of the interesting features is that the forces between particles depend strongly on the nature of the liquid, and so by changing the liquid we can exercise some control over the forces and hence over the stability and rheology of a powder dispersion.

The principal aim here will be to outline the various contributions to particle interactions in suspensions. Although two of the sections below contain a significant number of equations, the emphasis will be on describing the underlying mechanisms rather than giving rigorous theoretical accounts of the forces, in the hope that this chapter will impart some sense of the physical effects involved in suspensions. Those interested in more detail can consult recent texts, such as Israelachvili [2], which gives a very good introduction to surface forces in general; Hunter [3], which is a little more advanced and more directed towards colloid science; Russel, Saville and Schowalter [4], which is a sophisticated text giving more attention to dense dispersions and rheology; or Reed [5] for a comprehensive description of all the steps involved in ceramic processing. Only a few references to original work on surface forces are given in this chapter; a comprehensive list can be found in a recent review article [6].

3.2 SURFACE FORCES

The force between two bodies results from a summation of all of the **interatomic** forces acting between all of the atoms of the materials involved; those of the two bodies plus any intervening medium. Because of many-body effects (Fig. 3.1) the summation is not a simple pairwise addition, although some of the correct trends in behaviour can be understood by considering that as a first approximation.

The relevant interatomic forces can be classified according to their range:

1. Coulombic forces which have a long range,
2. van der Waals forces which are considered 'long-range' on an atomic scale but are short-range on a macroscopic scale, and
3. very-short-range forces resulting from the exchange or non-exchange of electrons: covalent bonding, hydrogen bonding and Born repulsions. (Metallic bonding will not be considered here.)

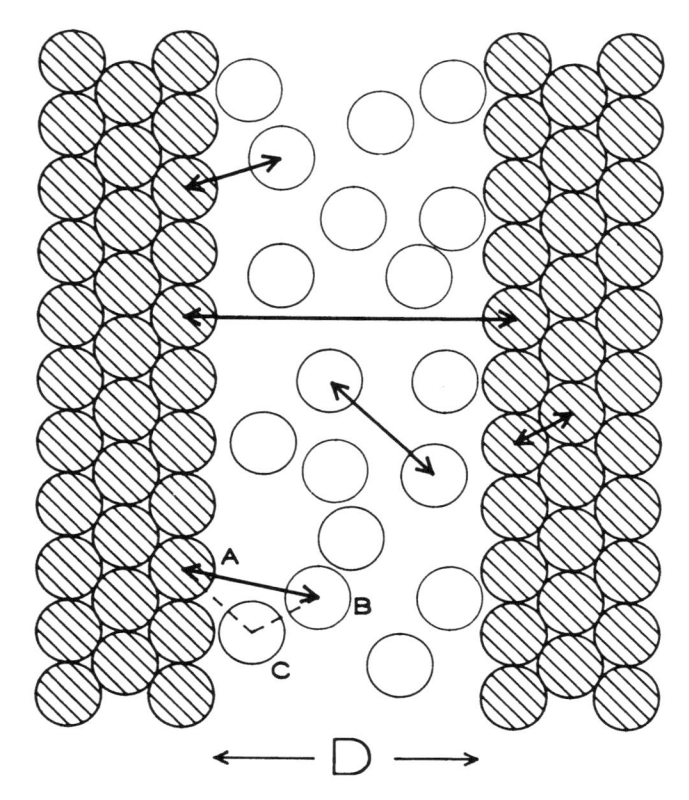

Fig. 3.1 The force between two bodies separated by a fluid medium results from the net effect of the forces between all of the atoms and molecules present; because of many-body effects (the force between A and B is affected by the proximity of C), the total is not a simple sum of all the pair interactions.

When all the interatomic forces are summed, the very-short-range forces remain very short-range, and contribute only to the interaction of atoms with their immediate neighbours. Thus it is only atoms at the surfaces or interfaces between materials that contribute to these forces, although as we shall see below, their effect can sometimes propagate for several atomic or molecular diameters in a liquid.

The van der Waals interactions decrease with the inverse seventh power of distance between atoms; summing over all the atoms in a body increases this range significantly, to the inverse third or second power depending on the geometry of the bodies. But it is still the closest atoms which dominate the summation, i.e. those nearest the surfaces of the bodies, and the resulting force is a function of the **surface** separation. Long-range Coulombic interactions arise between charges, and these most often reside at the surfaces of the bodies. Later we will encounter other contributions, such as capillary and solvation effects, arising from the nature of the medium between the bodies and its interaction with the surfaces. Forces due to these effects depend on the thickness of the medium. Thus all of these contributions to the force between two bodies depend not on the distance between the centres of the bodies but the distance between their **surfaces**, hence the name **surface forces**.

It is a common and convenient practice to assume that the total force between two surfaces is obtained simply by summing several different components. This approach, initially suggested by the leading Russian colloid scientist Boris Derjaguin, implicitly assumes that the components are independent of each other. Generally this turns out to be a fairly good approximation, although on occasions there is some subtle interplay between different components, or some difficulty in deciding what label to apply to a certain effect. Let us now go on to describe the various components of surface forces between solids immersed in a liquid.

3.2.1 Forces and energies

There is a quantity called the **interaction free energy per unit area**, E_a, which is the difference in free energies between a system having two planar surfaces immersed in a medium at a separation D and a system having the same two surfaces immersed in the same medium, but at a large separation so that the surfaces do not interact with each other. The interaction energy is equal to the work done against the surface forces (per unit area), F_a, in bringing the surfaces from 'infinite' separation to D

$$E_a(D) = \int_\infty^D F_a(D)\,dD, \tag{3.1}$$

and the surface force is the negative gradient of the interaction energy

$$F_a(D) = -\frac{dE_a(D)}{dD}. \tag{3.2}$$

In the discussions below it will sometimes be convenient to think of interaction energies and sometimes of forces, but the two are easily related to each other through equations (3.1) or (3.2).

3.2.2 Flat and curved surfaces: the Derjaguin approximation

There is a very useful relationship associated with the name of Derjaguin, which makes a correspondence between the **force**, F_c, between curved particle surfaces and the interaction **energy** per unit area between flat surfaces. Most theories of surface forces are worked out for flat surfaces, because this is by far the simplest geometry; but of course in the real world we want to know what the forces are between finite-sized objects, and these are rarely flat (the most common exception would be clay platelets). The Derjaguin approximation

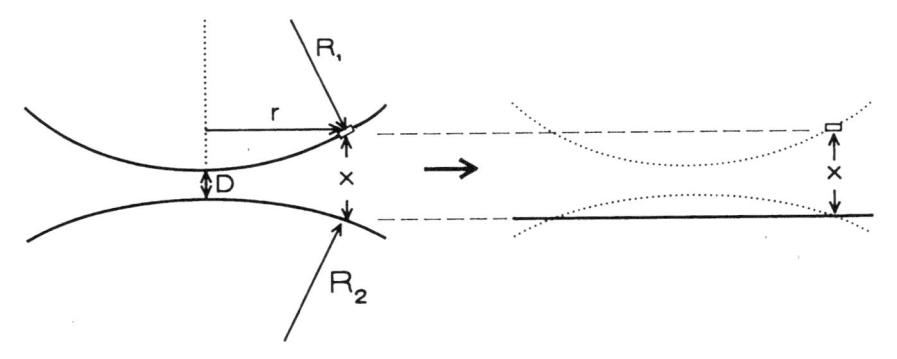

Fig. 3.2 In the Derjaguin approximation an element of curved surface is replaced by an element of flat surface interacting with a second flat surface at the separation of the element directly opposite.

allows us to compute the force between curved surfaces from a knowledge of the energy between flats, at least in some situations when certain conditions are satisfied.

To obtain the force between curved surfaces (e.g. two spheres of radii R_1 and R_2) we start by dividing one surface into incremental elements of area, then considering the interaction energies, E_{el}, between those elements and the other surface, then integrating over all elements to get the total interaction energy, E_c. The first approximation is to replace each curved surface element by an element of a **flat** surface lying parallel to the tangent plane at the point of closest approach. The interaction energy of an element is taken as being the same as the interaction energy between the element and a second flat surface at the separation x directly opposite (Fig. 3.2). In doing this, the energy is assumed to be independent of the **curvature** of the surfaces, which will be a reasonable approximation for small curvatures (large radii). Thus, integrating over the first surface S_1,

$$E_c = \int_{S_1} E_{el} \, dS$$

$$= \int_{S_1} E_a(x) \, dS$$

$$= \int_0^{2\pi} d\theta \int_0^{R_1} r \, dr E_a[x(r)]. \tag{3.3}$$

where θ is the azimuthal angle. The second approximation is that surface curvature is parabolic,

$$x \approx D + \frac{r^2}{2R_1} + \frac{r^2}{2R_2}, \tag{3.4}$$

giving

$$E_c(D) = 2\pi R \int_D^{R_1 + D} E_a(x)\,\mathrm{d}x, \tag{3.5}$$

where

$$R = \frac{R_1 R_2}{R_1 + R_2}. \tag{3.6}$$

The third approximation now comes in assuming that the **range** of the surface force is small compared to the radius of either particle, so that the upper limit to the integral in equation (3.5) can be replaced by infinity. Since (from equation (3.2)) the force between curved surfaces $F_c = -\mathrm{d}E_c/\mathrm{d}D$, taking the derivative of equation (3.5) leads to the Derjaguin approximation

$$F_c(D) = 2\pi R E_a(D). \tag{3.7}$$

3.2.3   van der Waals forces

Perhaps the most familiar force between particles is the van der Waals force. It is also one of the most important in determining the stability of dispersions, particularly the dense (high volume fraction) suspensions that are required for powder processing of ceramics.

The van der Waals force is ubiquitous, and is always attractive between like particles. It arises from the interaction of atomic and molecular electric **dipoles** whose orientations are **correlated** in such a way that they attract each other. Three types of interaction can be distinguished.

- A permanent molecular dipole creates an electric field which has the effect of orienting other permanent dipoles so that they are attracted to the first one. This is sometimes called the Keesom interaction.
- A permanent dipole **induces** a dipole in a polarizable atom, molecule or medium, and the induced dipole is oriented such that it is attracted; this is called the Debye interaction.
- An instantaneous dipole, arising from a fluctuation in the distribution of electronic charge, itself induces other dipoles in surrounding atoms and molecules. Once again, the induced dipoles are attracted to the inducing one. This is known as the London or dispersion force. In some disciplines the term 'van der Waals force' means only the London interaction; in others it is taken to include all three. If there are no permanent dipoles present in a system then the first two contributions do not occur, and there is no confusion. We will see below that there is still another effect which could reasonably be considered as a contribution to the van der Waals force; it arises from correlations in the positions of **charges**. Perhaps the definition should be broadened to include all attractive forces which are the result of **correlations** between distributions of charge, whether as dipoles or as isolated charges in, on, and between different media.

The Keesom, Debye and London forces exist between molecules, all varying as d^{-7}, where d is the intermolecular distance. The London force between atoms or molecules which are far apart is weaker, falling off as d^{-8}, due to an effect known as **retardation**. In that case, by the time the electric field has propagated at the speed of light from one instantaneous dipole, induced the second, and returned to the first, the first dipole will have rotated and no longer be correlated with the induced one. To calculate the distance dependence of the force between two macroscopic bodies, Hamaker and de Boer in 1937 made a pairwise summation over all the atoms in the bodies, and showed that for large bodies the force depends on their **surface** separation D. For example, the non-retarded force between two spheres of radius R_s is given by

$$F(D) = -\frac{AR_s}{12D^2} \tag{3.8}$$

where A, the **Hamaker constant**, depends on the polarizabilities and number densities of the atoms in the two bodies. This approach is illuminating in that it illustrates how summing the forces between atoms significantly increases the **range** of interaction between particles. However, the Hamaker approach turns out to be not entirely accurate because of many-body effects: the fact that the field which one atomic dipole creates at a neighbour is also influenced by other atoms in the vicinity, so that the total force is not obtained by simply adding up the effects of each pair of atoms.

An alternative approach, due to Lifshitz, is to consider each body as a **continuum** with certain dielectric properties; these automatically incorporate many-body effects. Then, using quantum field theory, one sums over all fluctuation modes in the electromagnetic field, and finds an energy which decreases as two bodies approach, that is, an attractive force. Permanent dipoles (Keesom and Debye contributions) are incorporated in a so-called **zero-frequency** term. The full theory is very complicated. Fortunately, several authors have set out good descriptions of the Lifshitz theory in far more comprehensible (though still not trivial) terms [7–9].

The van der Waals force in the continuum Lifshitz approach depends on the dielectric properties of each medium at all frequencies ω (rad s^{-1}). These are described by complex dielectric response functions

$$\varepsilon(\omega) = \varepsilon'(\omega) + i\varepsilon''(\omega). \tag{3.9}$$

At zero frequency the real part ε' is proportional to the familiar dielectric constant (or relative dielectric permittivity), ε_r

$$\varepsilon'(0) = \varepsilon_0\varepsilon_r, \tag{3.10}$$

while at optical frequencies ω_{opt} it is proportional to the square of the refractive index n

$$\varepsilon'(\omega) = \varepsilon_0 n^2, \tag{3.11}$$

where ε_0 is the permittivity of free space. The complex part ε'' gives the

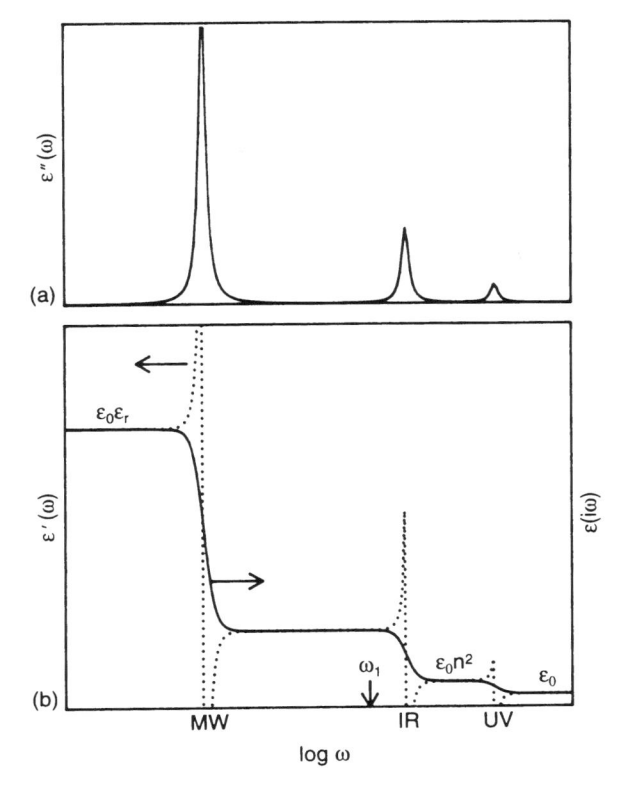

Fig. 3.3 A schematic diagram of dielectric response functions as a function of frequency. Part (a) shows the imaginary component, $\varepsilon''(\omega)$, corresponding to the absorption spectrum, of a notional material having absorption bands in the microwave (MW), infrared (IR) and ultraviolet (UV) regions. Part (b) shows two related functions: the real component $\varepsilon'(\omega)$ (dotted line) and the quantity $\varepsilon(i\omega)$ (solid line). When divided by the permittivity of free space ε_0, these quantities have the value of the static dielectric constant ε_r at low frequency, the refractive index n^2 at optical frequencies, and unity at infinite frequency. Both $\varepsilon'(\omega)$ and $\varepsilon(i\omega)$ are related to $\varepsilon''(\omega)$ through similar Kramers–Kronig relations (equations (3.12) and (3.19)); $\varepsilon(i\omega)$ is a well-behaved, monotonically decreasing function which is essentially equal to the more familiar $\varepsilon'(\omega)$ in the regions between absorption frequencies.

dielectric **loss**, or absorption spectrum of the material; it is zero in frequency ranges over which the material is transparent (Fig. 3.3).

An expression known as a Kramers–Kronig relation links the real and imaginary parts of the dielectric function

$$\varepsilon'(\omega) = \varepsilon_0 + \frac{2}{\pi} \int_0^\infty \frac{v\varepsilon''(v)}{v^2 - \omega^2} \, \mathrm{d}v . \qquad (3.12)$$

The result of the Lifshitz theory for the interaction free **energy** per unit area between two halfspaces 1 and 2 separated by a thickness D of medium

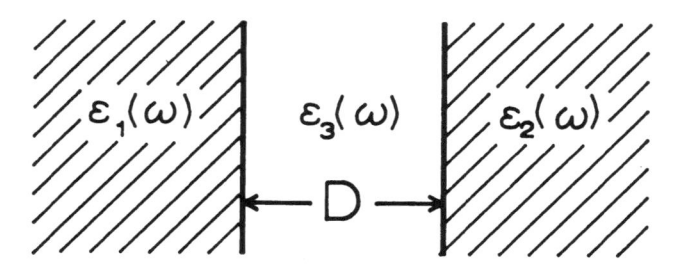

Fig. 3.4 Two dielectric halfspaces 1 and 2 separated by a thickness D of medium 3. The materials are characterized by their dielectric response functions $\varepsilon_j(\omega)$.

3 (Fig. 3.4) in the non-retarded regime, i.e. at small separations (< 5 nm), is

$$E_{1\,3\,2}(D) = -\frac{A_{1\,3\,2}}{12\pi D^2} \tag{3.13}$$

where

$$A_{1\,3\,2} = -\frac{3kT}{2} \sum_{m=0}^{\infty}{}' \int_0^{\infty} \mathrm{d}x\, x \ln[1 - \Delta_{1\,3}\Delta_{2\,3}e^{-x}] \tag{3.14}$$

is the Hamaker constant. Here k is the Boltzmann constant, T is the absolute temperature, and the prime on the summation indicates that half-weight is given to the $m = 0$ term. The Hamaker constant depends on the dielectric response functions $\varepsilon_j(\omega)$ of the three media $j, k = 1, 2, 3$ through the expressions

$$\Delta_{jk} = \frac{\varepsilon_j(i\omega_m) - \varepsilon_k(i\omega_m)}{\varepsilon_j(i\omega_m) + \varepsilon_k(i\omega_m)} \tag{3.15}$$

with

$$\omega_m = m\left[\frac{4\pi^2 kT}{h}\right], \tag{3.16}$$

where h is Planck's constant and m is an integer.

The integral in equation (3.14) gives

$$A_{1\,3\,2} = \frac{3kT}{2} \sum_{m=0}^{\infty}{}' \sum_{s=1}^{\infty} \frac{(\Delta_{1\,3}\Delta_{2\,3})^s}{s^3}. \tag{3.17}$$

It can be shown that the $s = 1$ term is by far the most important, and the summation in s can be truncated after one term with an error that is usually less than 5%. Thus

$$A_{1\,3\,2} \approx \frac{3kT}{2} \sum_{m=0}^{\infty}{}' (\Delta_{1\,3}\Delta_{2\,3}) \tag{3.18}$$

In equation (3.15), rather mysterious quantities appear: the terms $\varepsilon(i\omega)$ are the dielectric responses at **imaginary** frequencies! Do not be alarmed by the apparently unphysical nature of $\varepsilon(i\omega)$; this is simply a convenient mathematical trick. There is a second Kramers–Kronig relation which brings us back to the real world

$$\varepsilon(i\omega) = \varepsilon_0 + \frac{2}{\pi} \int_0^\infty \frac{\nu \varepsilon''(\nu)}{\nu^2 + \omega^2} \, d\nu. \qquad (3.19)$$

The similarity between equations (3.12) and (3.19) shows that $\varepsilon(i\omega)$ is a function with some resemblance to $\varepsilon'(\omega)$, as illustrated in Fig. 3.3. If the complete dielectric loss spectrum $\varepsilon''(\omega)$ is known for all of the materials involved, then the Hamaker constant (and the van der Waals force) can be computed very accurately.

More commonly, the entire absorption spectrum is not known, and approximate methods of interpolation or extrapolation from known experimental data are required to estimate the Hamaker constant. In some cases these can still be quite accurate, particularly when there is only a small number of absorption bands, and when reasonable data are available in those regions.

It turns out that the ultraviolet region of frequencies is particularly important in evaluating $\varepsilon(i\omega)$. This is because we require the values at a series of frequencies given by equation (3.16), i.e. multiples of $(4\pi^2 kT/h)$, which is 2.5×10^{14} rad s^{-1} at 300 K. Therefore there are no sampling points in the microwave region and only a few in the infrared region ($< 3 \times 10^{15}$ rad s^{-1}), but there are several hundred through the visible and ultraviolet where $\varepsilon(i\omega)$ may still be reasonably large. For many transparent materials there is a single absorption peak in the ultraviolet region, which can be characterized quite accurately by extrapolation from visible refractive index data. One way to do this [9] is by means of a Cauchy representation

$$\varepsilon'(\omega) = \varepsilon_0 n^2(\omega) = \varepsilon_0 \left(1 + \frac{C_{UV}}{1 - (\omega/\omega_{UV})^2} \right), \qquad (3.20)$$

where C_{UV} is the **oscillator strength** of the absorption at frequency ω_{UV}. By rearranging equation (3.20) it can be shown that refractive index data in the visible region plotted as $[n^2 - 1]$ against $[n^2 - 1]\omega^2$ should give a straight line of slope $1/\omega_{UV}^2$ and intercept C_{UV}. The requisite dielectric function at imaginary frequency is given by

$$\varepsilon(i\omega) = \varepsilon_0 \left(1 + \frac{C_{UV}}{1 + (\omega/\omega_{UV})^2} \right). \qquad (3.21)$$

Putting these values into equations (3.15) and (3.18) would give an accurate estimate for the Hamaker constant when suitable refractive index data are available. (An alternative formulation can be obtained based on the more familiar Clausius–Mosotti plot, leading to an equivalent form for $\varepsilon(i\omega)$ with slightly different parameters [9].)

If all three materials $i = 1, 2, 3$ happened to have the same UV absorption frequency ω_{UV} the expression for the Hamaker constant could be expressed in terms of the refractive indices n_i

$$A_{132} \approx \frac{3kT}{2} \sum_{m=0}^{\infty}{}' \frac{(n_1^2 - n_3^2)(n_2^2 - n_3^2)}{[n_1^2 + n_3^2 - (1 - 3b_m^2)/(1 - b_m^2)][n_2^2 + n_3^2 - (1 - 3b_m^2)/(1 - b_m^2)]},$$

(3.22)

where

$$b_m = \frac{\omega_m}{\omega_{UV}} = \frac{4\pi^2 kTm}{h\omega_{UV}},$$

(3.23)

and the refractive indices in each term of the sum in equation (3.22) would be the values taken at the frequency ω_m. Such a coincidence of absorption frequencies is not very likely to occur in practice, but the simplification serves to illustrate two important features of the van der Waals force.

• The Hamaker constant depends primarily on the **differences** in refractive indices between the two solid half spaces (1, 2) and the (fluid) medium (3) separating them. Thus the magnitude of the van der Waals force can be minimized by 'index matching', if the dielectric properties of the intervening medium can be made similar to those of one or other of the solids.
• If it should happen that $n_1 > n_3 > n_2$, the sign of the Hamaker constant becomes negative. This means that the van der Waals force between unlike materials can be a **repulsion** if the dielectric properties of the intervening medium are intermediate between those of the other two. However, the force between **like** materials (1 = 2) is always attractive. The van der Waals repulsion is more than just a curiosity; it can be significant in certain situations such as the wetting or de-wetting of hydrocarbon films on the surface of water, but it is unlikely to occur in ceramic processing systems.

3.2.4 Electrostatic forces in a non-polar environment

Surfaces can become charged either by adsorbing or desorbing electrons or ions, and they then interact with each other via a direct Coulombic force. It is reasonable to suppose that the charging process depends on the surface properties and on the environment, so that particles of the same material are likely to acquire the same charge and thus repel each other, but they may be attracted to a different material. Because of image charge effects a charged particle will also be attracted to an uncharged material whose dielectric constant is higher than that of the surrounding medium.

The direct Coulombic interaction is generally stronger and longer-ranged than other surface forces, so when it occurs it will be the major determinant of the behaviour of a system. This is well known for powders in air. The same kind of interactions could occur in non-polar liquids, but much less is known about that situation. It is known that particles can be electrostatically stabilized in non-polar liquids, at least at low volume fraction. The long-range nature of the force would make it difficult to describe a system of charged particles at high volume fraction, because many-body effects could not be ignored. Another complication is that one should consider what happened to the charges of opposite sign which were presumably generated when the surfaces first became charged. Do they remain in the liquid? If so, then the situation begins to resemble that outlined in the following section, although the concept of electrical double layers may not hold much meaning in a non-polar liquid where the solubility of ions is very low.

3.2.5 Electrostatic forces in a polar liquid: double-layer repulsion

When a solid is immersed in a polar solvent such as water, it usually acquires a surface charge, either by adsorbing or desorbing ions according to some chemical equilibrium with the surrounding solution. For example, many oxide surfaces such as clays, silica (and the native oxide films on silicon nitride and silicon carbide), alumina, titania, zirconia, etc., have surface groups which are hydroxylated, and undergo proton association– dissociation reactions of the form

$$(\text{low pH}) \; -M^+ - OH_2 \overset{H^+}{\longleftarrow} \; -M - OH \overset{OH^-}{\longrightarrow} \; -M - O^- + H_2O \; (\text{high pH}),$$

$$(3.24)$$

where $-M$ represents the metal (or silicon) atom. Thus the surface is positively charged at low pH and becomes negative at high pH (Fig. 3.5). There is a certain pH, called the point of zero charge (PZC) (or the isoelectric point

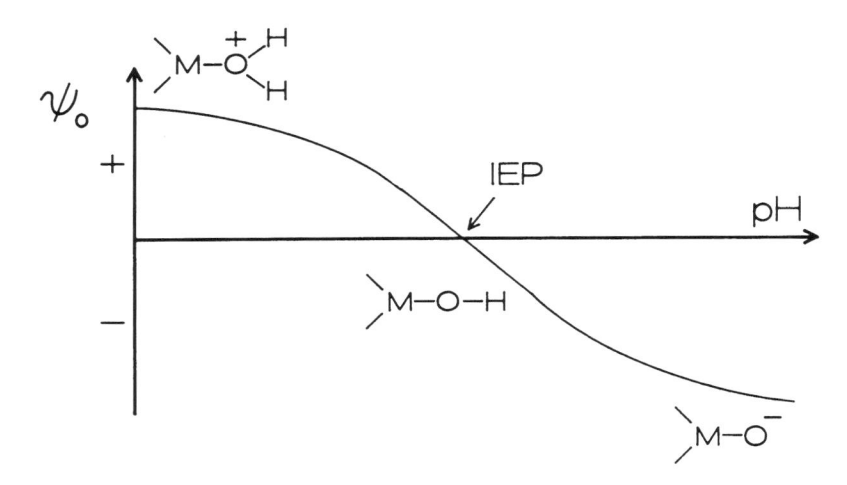

Fig. 3.5 Typical variation of surface potential ψ_0 with pH for a metal oxide surface, as measured for example by electrokinetic techniques. Surface charge has the same sign as surface potential; at the isoelectric point (IEP) the surface has no net charge.

(IEP) if determined from electrokinetic measurements) at which the total charge on the surface is zero and the electrical repulsion between two such surfaces (discussed below) is eliminated.

Ions of opposite charge which are dissolved in the polar medium, known as **counterions**, are attracted towards the surface. However, they do not simply stick to the surface. As a consequence of the balance between their electrostatic energy (favouring counterions being close to the surface) and entropic considerations (which favour them spreading throughout the liquid), these ions form a **diffuse** layer of charge adjacent to the surface. The surface charge plus the diffuse layer of opposite charge constitute an **electrical double layer**.

Descriptions of the electrical double layer can be found in standard texts [2–4]; here we follow the treatment given by Hunter [3]. The basic equation is obtained by combining Poisson's equation relating electrical potential ψ to charge density ρ

$$\nabla^2 \psi = -\frac{\rho}{\varepsilon_0 \varepsilon_r}, \tag{3.25}$$

with the Boltzmann equation giving the distribution of ions according to their electrostatic energy

$$N_i = N_i^0 \exp(-z_i e\psi/kT) \tag{3.26}$$

Here N_i is the local number density and N_i^0 is the bulk concentration of ion species i having valency z_i, and e is the electronic charge. Since the charge density is given by

$$\rho = \sum_i N_i z_i e, \tag{3.27}$$

Equations (3.25) and (3.26) can be combined to give the Poisson–Boltzmann equation

$$\nabla^2 \psi = \frac{-1}{\varepsilon_0 \varepsilon_r} \sum_i N_i^0 z_i e \exp\left(\frac{-z_i e\psi}{kT}\right). \tag{3.28}$$

For a planar surface this reduces to a one-dimensional equation in x (the distance from the surface). For a symmetrical $z:z$ electrolyte at bulk concentration N^0 (far from the surface) we obtain

$$\frac{d^2\psi}{dx^2} = \frac{2N^0 ze}{\varepsilon} \sinh\left(\frac{ze\psi}{kT}\right), \tag{3.29}$$

in which we have put $\varepsilon = \varepsilon_0 \varepsilon_r$. Multiplying both sides of equation (3.29) by $2d\psi/dx$ and taking the integral with respect to x,

$$\frac{2d\psi}{dx}\frac{d^2\psi}{dx^2} = \frac{4N^0 ze}{\varepsilon} \sinh\left(\frac{ze\psi}{kT}\right)\frac{d\psi}{dx} \tag{3.30}$$

$$\int \frac{d}{dx}\left(\frac{d\psi}{dx}\right)^2 dx = \int \frac{4N^0 ze}{\varepsilon} \sinh\left(\frac{ze\psi}{kT}\right)d\psi. \tag{3.31}$$

Integrating from ∞ (where $\psi = 0$ and $d\psi/dx = 0$) to a distance x gives

$$\left(\frac{d\psi}{dx}\right)^2 = \frac{4N^0kT}{\varepsilon}\left[\cosh\left(\frac{ze\psi}{kT}\right) - 1\right], \quad (3.32)$$

this reduces to

$$\frac{d\psi}{dx} = -\frac{2\kappa kT}{ze}\sinh\left(\frac{ze\psi}{2kT}\right), \quad (3.33)$$

where we have introduced the **Debye constant**

$$\kappa = \left(\frac{e^2 \sum N_i^0 z_i^2}{\varepsilon kT}\right)^{1/2}. \quad (3.34)$$

Equation (3.33) can be integrated from the bulk solution up to the surface at $x = 0$ to give

$$\tanh\left(\frac{ze\psi}{4kT}\right) = \tanh\left(\frac{ze\psi_0}{4kT}\right)\exp(-\kappa x), \quad (3.35)$$

where ψ_0 is the **surface potential**, i.e. the potential at $x = 0$.

For low potentials (always the case far from the surface) the Poisson–Boltzmann equation can be linearized to

$$\psi = \frac{4kT}{ze}Z\exp(-\kappa x), \quad (3.36)$$

where

$$Z = \tanh\left(\frac{ze\psi_0}{4kT}\right). \quad (3.37)$$

In other words, to a first approximation the potential decreases **exponentially** with distance away from the surface, with a decay length (the **Debye length**) which is the inverse of the Debye constant. Thus the Debye length gives a measure of the thickness of the diffuse double layer. From equation (3.34) we see that it depends on the concentration of ions in solution: more ions available give a thinner double layer. In terms of the ionic strength $I = \frac{1}{2}\sum c_i z_i^2$, where c_i is the ionic concentration in mol l^{-1}, the Debye length (in m) is

$$\kappa^{-1} = \left(\frac{2000e^2 N_A I}{\varepsilon kT}\right)^{-1/2}, \quad (3.38)$$

where N_A is Avogadro's number. In water at 25 °C the Debye length in nm is $0.304/I^{1/2}$, which for 10^{-3} M 1:1 electrolyte gives $\kappa^{-1} = 9.6$ nm.

The total charge (per unit area of surface) in the diffuse layer is

$$\sigma_d = \int_0^\infty \rho \, dx. \quad (3.39)$$

This charge must balance the **surface charge** per unit area σ_0. The integral in equation (3.39), using equations (3.25) and (3.33), relates surface charge to surface potential

$$\sigma_0 = -\varepsilon \left(\frac{\mathrm{d}\psi}{\mathrm{d}x} \right)_{x=0}$$

$$= \frac{2\kappa kT\varepsilon}{ze} \sinh \left(\frac{ze\psi_0}{2kT} \right) \tag{3.40}$$

For a symmetrical electrolyte at concentration $c \, \mathrm{mol}\,l^{-1}$ in water at $25\,°\mathrm{C}$,

$$\sigma_0 = 0.1174 c^{1/2} \sinh (19.46 z\psi_0) \tag{3.41}$$

in $C\,m^{-2}$ when ψ_0 is in volts.

At this point we should note that there are various approximations implicit in the derivation of the Poisson–Boltzmann equation. In particular, the counterions are assumed to be point charges, the solvent is assumed to be a structureless continuum characterized only by its dielectric constant, and the surface charge is assumed to be smeared out uniformly over the surface. Furthermore, in the simple treatment given here we have taken the surface charge and surface potential to be defined at the plane $x = 0$. In doing so we have avoided the detailed issues of exactly how charges are distributed near a surface, whether bound or adsorbed charges are excluded from the diffuse layer, and so on. These considerations result in terms such as Stern layer, inner layer, Helmholtz plane, triple layer and others. We will not endeavour to discuss any of those models here, because our interest is in the repulsion between two charged surfaces, and that is determined primarily by the diffuse layer. It should be noted, however, that the surface charge as we have defined it may not correspond to the surface charge discussed in other models. For example, it may be possible to measure a surface charge by titration experiments (equation (3.24)), but the titratable charge is not necessarily the same thing as the double-layer surface charge σ_0.

Particles of the same material would acquire the same charge on immersion in water, and consequently repel each other. However, the magnitude and range of the repulsion are significantly reduced compared to the direct Coulombic repulsion which would occur between the bare surface charges alone. There are different ways to think of the interaction between charged surfaces in a polar liquid. One is to say that the presence of counterions in the diffuse layer **screens** the direct Coulombic interaction. Another is to consider the change (an increase) in free energy of the system when two diffuse double layers begin to **overlap**, so that their combined structure is distorted from the structure of two isolated double layers (described above). A third is to note that when the diffuse layers approach sufficiently closely that they overlap, the concentration of counterions midway between the surfaces is increased over the background concentration, which gives rise to an excess osmotic pressure. The increased osmotic pressure equals the mechanical pressure required to hold two surfaces at that separation.

Any of these views would lead to the realization that the **range** of the force between two such surfaces is of the order of a few times the Debye length. The

double-layer repulsion (as it is called) is long-range at low electrolyte concentration (perhaps 100 nm in distilled water) and short-range in concentrated electrolytes (e.g. ~ 1 nm at 0.1 м).

The Poisson–Boltzmann equation (3.29) still holds when two surfaces approach so that their diffuse double layers overlap (Fig. 3.6). Following a similar procedure to equations (3.30)–(3.32) gives

$$2N^0 kT \cosh\left(\frac{ze\psi}{kT}\right) - \frac{\varepsilon}{2}\left(\frac{d\psi}{dx}\right)^2 = p_m, \qquad (3.42)$$

where p_m is a constant. The first term in this equation is just the osmotic pressure, since

$$2N^0 kT \cosh\left(\frac{ze\psi}{kT}\right) = kT\left[N^0 \exp\left(\frac{ze\psi}{kT}\right) + N^0 \exp\left(-\frac{ze\psi}{kT}\right)\right]$$

$$= kT(N^+ + N^-), \qquad (3.43)$$

N^+ and N^- being the number densities of cations and anions. The second term

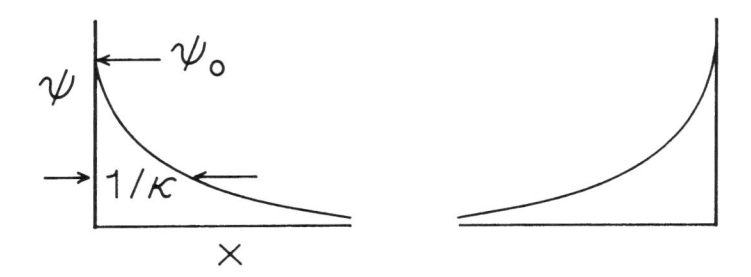

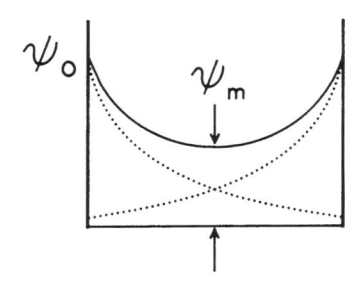

Fig. 3.6 (Top) Electrostatic potential ψ in the double-layer adjacent to isolated charged surfaces in a polar liquid such as water. Potential decreases quasi-exponentially away from the surfaces, with a decay length given by the reciprocal of the Debye constant (equation (3.34)). (Bottom) When two surfaces approach, their double-layers overlap and the potential profile is altered, although potential still obeys the Poisson–Boltzmann equation (3.29) (with different boundary conditions). The double-layer force between the two surfaces is related to the mid-plane potential ψ_m.

in equation (3.42) is zero at the midplane between the two planar surfaces where, by symmetry, $d\psi/dx = 0$. Hence the constant p_m is just the osmotic pressure at the midplane. But as we noted above, the mechanical pressure (or force per unit area F_a) between planar surfaces is equal to the increase in osmotic pressure, i.e. the difference in osmotic pressures between the midplane and the bulk liquid. Thus the double-layer repulsion is given by

$$F_a = kT(N^+ + N^- - 2N^0)$$

$$= 2N^0kT\left[\cosh\left(\frac{ze\psi_m}{kT}\right) - 1\right]. \tag{3.44}$$

where ψ_m is the potential at the midplane (Fig. 3.6). This is not all that useful, since ψ_m is not generally known nor is it easily measurable; what we really want is the force in terms of the surface charge or the surface potential, since these quantities are more accessible experimentally.

There are methods of computing ψ_m and hence F_a by solving the Poisson–Boltzmann equation numerically [10]. We will not attempt to describe those methods here, but will give a simpler solution for the linearized Poisson–Boltzmann equation (equation (3.36)), which is valid for low surface potentials ($< 50\,mV$) and 'weak overlap', i.e. surfaces separated by more than a Debye length. In that case the midplane potential is given approximately by the sum of the potentials due to each surface, so for surfaces separated a distance D, we have (from equation (3.36))

$$\psi_m = \frac{8kT}{ze}Z\exp(-\kappa D/2). \tag{3.45}$$

The midplane potential is small enough to expand the cosh term in equation (3.44) to give

$$F_a = \frac{\kappa^2\varepsilon}{2}\psi_m^2$$

$$= 64N^0kTZ^2\exp(-\kappa D). \tag{3.46}$$

Although this simple expression for the repulsion between charged surfaces in a polar liquid is only valid under the conditions specified above, it does illustrate that to a first approximation the double-layer repulsion is an exponential function of surface separation, having a decay length equal to the Debye length.

As noted above, there are various simplifying assumptions inherent in the Poisson–Boltzmann equation. Current theoretical work is aimed at going beyond these assumptions, and there have been steady advances in obtaining a more precise picture of the double-layer. Broadly speaking, the approximations made in deriving the Poisson–Boltzmann equation will be good for surface separations greater than one Debye length; at shorter distances it may become important to consider the additional features included in modern theories. It has been shown theoretically that some of the improvements tend to cancel each other out, and experimentally that the Poisson–Boltzmann equation works very well down to separations of a few nanometres. At still smaller separations other effects take over; they will be discussed below.

3.2.6 DLVO theory

In the 1940s, the Russian scientists Derjaguin and Landau and the Dutch scientists Verwey and Overbeek independently suggested that the interparticle forces determining colloidal stability were given by the sum of van der Waals attraction and double-layer repulsion. The resultant theory, now known as DLVO theory, is the fundamental theory of colloid science. This combination of a positive (repulsive) exponential function of separation, whose range depends on ionic strength (equation (3.46)), with a negative (attractive) power-law function which is insensitive to ionic strength (equation (3.13)), gives rise to some interesting behaviour, as illustrated in Fig. 3.7. At small and at large separations the power law always dominates (although at large separations both terms may be imperceptibly small), while at intermediate separations the exponential repulsion usually exceeds the attraction. Thus the typical situation is one in which an energy barrier resulting from the repulsive force resists the approach of two particles, but if the particles collide with sufficient kinetic energy to overcome that barrier, the attractive force will pull them into contact ($D = 0$) where they adhere strongly (and irreversibly) to each other. There is also the interesting possibility of a 'secondary minimum' at finite separation (Fig. 3.7b) in concentrated electrolyte solutions. This gives a much weaker and potentially reversible adhesion between particles.

The range of the double-layer repulsion is easily controlled by changing the ionic strength of the solution in which particles are suspended, and, if the charging mechanism is understood, it is often possible to vary the strength of the repulsion too. For example, with oxide surfaces in water whose charge is determined by proton association–dissociation with surface hydroxyl groups, changing the pH will change the strength of the repulsion. At the IEP the repulsion disappears altogether. It is not so easy to control the van der Waals attraction, though there are tricks which may help, such as coating the particles with a material (e.g. a surfactant) of different dielectric properties. However, surfactant adsorption may also affect the forces in other ways, to be discussed below.

The DLVO theory has enjoyed considerable success in its original aim of predicting the stability of suspensions. It is further supported by direct measurements of the force between smooth surfaces of mica in water [11, 12] (Fig. 3.8) and other polar liquids, mica coated with surfactant, and surfaces of sapphire [13] and silica [14]. These measurements show that the range of the double-layer force matches that expected from the known electrolyte concentration in the solvent, and its magnitude is consistent with reasonable values of surface potential. In some cases the van der Waals attraction dominates when the surfaces are very close together, as predicted, but there are other cases in which it does not.

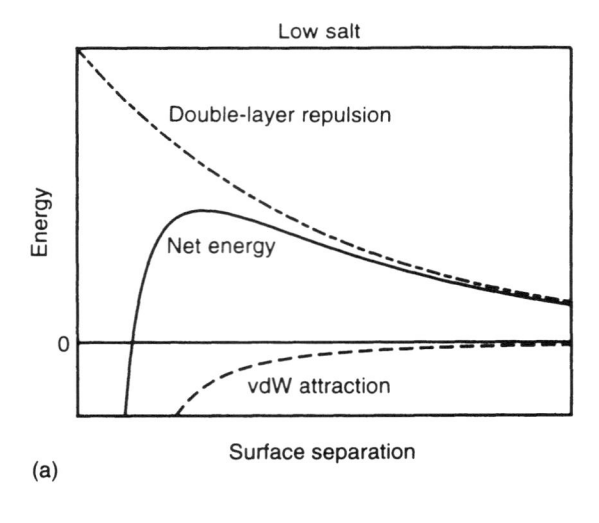

(a)

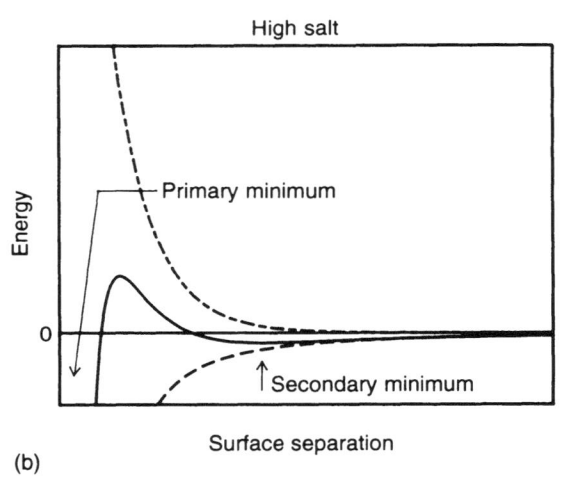

(b)

Fig. 3.7 Schematic diagram of the variation of interaction free energy with surface separation according to DLVO theory. The net energy (———) is given by the sum of a quasiexponential double-layer repulsion (–·–·) and a power-law van der Waals attraction (- - - - -). (a) With low salt concentration the repulsive term usually dominates at large separation, while the attractive term always dominates at short range, with an energy barrier in between (typically at 1–2 nm). (b) At high salt concentration the range of the repulsive term is reduced, so that the attractive term may again dominate at larger separations, giving a so-called secondary minimum when the surfaces are a few nanometres apart.

The latter situation illustrates a general limitation of the DLVO theory, namely that it does not work well when the colloidal particles are significantly solvated by the liquid in which they are suspended, as we shall see shortly (section 3.2.9). There are also difficulties in applying the theory to dense

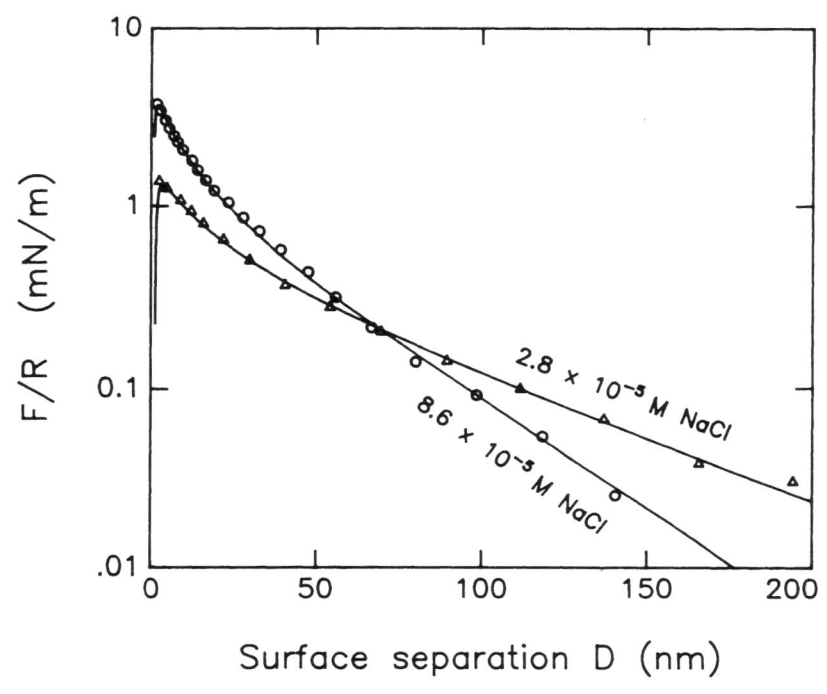

Fig. 3.8 Comparison between experimental and theoretical DLVO forces between two smooth sheets of mica immersed in water at low electrolyte concentration. Data (open circles and triangles) were obtained using an apparatus developed by Israelachvili [11] in which mica sheets are mounted as crossed cylinders. The measured force F is normalized by the radius R of the cylinders to relate these results to the energy between flat plates (equation 3.7). Solid lines are forces calculated from numerical solutions to the non-linear Poisson–Boltzmann equation [10]. The range of the double-layer force decreases as salt concentration increases, as expected. There is a maximum in the force at $D \approx 2\,\text{nm}$ and a primary minimum giving an adhesion between the surfaces at contact ($D = 0$).

suspensions, although it can be helpful in predicting the correct trends if not the quantitative behaviour.

The van der Waals and electrical double-layer forces which constitute DLVO theory are probably the best understood of all the surface forces, and it is partly for this reason that we have gone into some mathematical detail in describing them. The remainder of the forces discussed below are either less well quantified theoretically, or (like the following one) rather esoteric, and we will not attempt to present equations for them.

3.2.7 Charge correlation effects

DLVO theory assumes that the force due to permanent charges and the force due to instantaneous dipoles (high-frequency fluctuations of electronic charge) are quite independent and hence additive. The former is repulsive and is screened by counterions so its range depends on ionic strength; the latter is attractive, and insensitive to ionic strength. There is, however, a hazy area between these two: how do we consider the possible contribution from permanent dipoles whose orientations correlate, giving rise to a Keesom force, or which induce other dipoles and give a Debye force? Furthermore, recent theoretical work has demonstrated that non-uniform **charge** distributions in interacting double layers can also correlate with each other, so that when a charge of one sign occurs on one side of the midplane which divides the space between two surfaces, there is less likelihood of finding a charge of the same sign directly opposite it on the other side. This gives an additional **attractive** contribution to the force, or, equivalently, has the effect of reducing the usual double-layer repulsion. Such charge correlations are significant when the surfaces are highly charged and the counterions are polyvalent. The reduced repulsion which results from it has been observed experimentally, but it would not usually be an important effect in particle suspensions.

Because these effects arise from **correlations** and because they all give attractive contributions, it would seem logical to include them with the van der Waals force. However, because they all involve permanent charges (either isolated as ions or paired as dipoles), they contribute to the **static** electric field, which is screened by the counterions. (This appears in the Lifshitz theory of van der Waals forces as the zero-frequency term.) Thus the **range** of these force components also decreases as ionic strength increases, so perhaps they should be lumped together with the other electrostatic term: the double-layer repulsion. Evidently the usual division into an electrolyte-dependent repulsion and an electrolyte-independent attraction is not so clear-cut when all of the above possibilities are included.

3.2.8   Structural forces

In all of the forces described so far, the liquid separating the surfaces has been treated as a structureless continuum, which is a perfectly adequate approximation for large surface separations. However, in the real world everything consists of atoms and molecules, and if we work on a sufficiently fine scale we will have to be concerned with the discreteness of matter. When we do this we encounter a **structural force**, so called because it is a consequence of the arrangement of molecules, or **structure**, of the medium between the surfaces.

When a simple liquid of spherical molecules is confined to a narrow region between two flat plates, the gap imposes constraints on the liquid structure. The molecules prefer to order in **layers** parallel to the surfaces because this enables them to pack more efficiently; when the width of the gap does not allow an integral number of layers, the packing is inefficient and as a consequence the free energy of the liquid is increased. This leads to the prediction that the force between two smooth solid surfaces separated by such a liquid should show alternating maxima and minima as surface separation varies, with the amplitude of the extrema diminishing with increasing gap width until the layering effect of the surfaces is no longer significant, and the system goes over to continuum behaviour. The spacing between successive maxima (or minima) should be approximately equal to the molecular diameter.

This force has been observed between smooth mica surfaces separated by an inert, non-polar liquid. The ordering of molecules into layers is only expected to occur next to a surface that is rigid and smooth; the 'oscillations' will be smoothed out when the surface roughness exceeds the size of the liquid molecules. For this reason it probably does not have great significance for suspension stability or rheology, though it may be important in determining the thickness of intergranular phases after liquid phase sintering, and in certain questions of crack propagation in ceramics.

3.2.9 Solvation forces

When surfaces are immersed in a liquid, the force between them is often greatly affected by the interaction of that liquid with the surface. The surface may be **solvated** in a particular way; for example, a polar surface may orient the molecules of a polar liquid such as water, or the liquid molecules may hydrogen bond to an appropriate surface. An isolated surface will thereby modify the structure of the liquid adjacent to it (and the liquid may in turn modify the structure of the solid), with the modified structure extending for some distance – the thickness of some notional 'solvation layer' – into the liquid. The nature of the solvation layer depends on the strength and type of the perturbation and its thickness depends on how that perturbation is propagated through the liquid. Generally, it results in a region where the liquid structure is inhomogeneous, due to the competing requirements of the surface and the bulk liquid. When two solvated surfaces come close enough together so that the solvation layers overlap, the total energy of the system begins to change as a function of separation, which means that there is a force between the surfaces. (This language is reminiscent of that used to describe the repulsion resulting from the overlap of two diffuse double layers in section 3.2.5. In fact, the familiar electrical double-layer repulsion could also be thought of as a 'solvation force', where the term 'solvation' would refer to the local charge density.)

Our knowledge of solvation forces is based almost entirely on experimental evidence. Following some early work in Russia, a notable demonstration of the existence of solvation forces came in the measurement of a repulsion between uncharged bilayers of lecithin, a common biological lipid, immersed in water [15]. The force extends about 3 nm, or about ten water molecule diameters, and is generally explained in terms of a perturbed water structure extending from the hydrated surfaces. It was dubbed a **hydration force**, the particular name for a solvation force in water. Similar hydration forces have since been measured between mica surfaces [11, 12], silica surfaces [14], and bilayers of other (non-biological) surfactants. When present, the hydration force is a strong repulsion which exceeds the van der Waals attraction, thereby preventing surfaces from coming into a primary minimum at contact. Thus strongly solvated particles may form suspensions that are more stable than the DLVO theory would predict.

There is no general theory of solvation forces, because such forces depend on specific features of the particular surface and solvent involved. Some attempts have been made to offer theoretical explanations for the observed experimental effects, but it is fair to say that there is currently no theory capable of **predicting** whether a solvation force will occur and what its magnitude would be for a particular solid–liquid combination.

3.2.10 'Hydrophobic' forces

Recently, experiments have revealed the existence of an important and dramatic new force. It is a strong, long-range attraction which occurs between macroscopic hydrophobic surfaces immersed in water, and has been called the **hydrophobic force**. First measured directly with a mica surface made hydrophobic by adsorption of a monolayer of cationic surfactant, its existence has

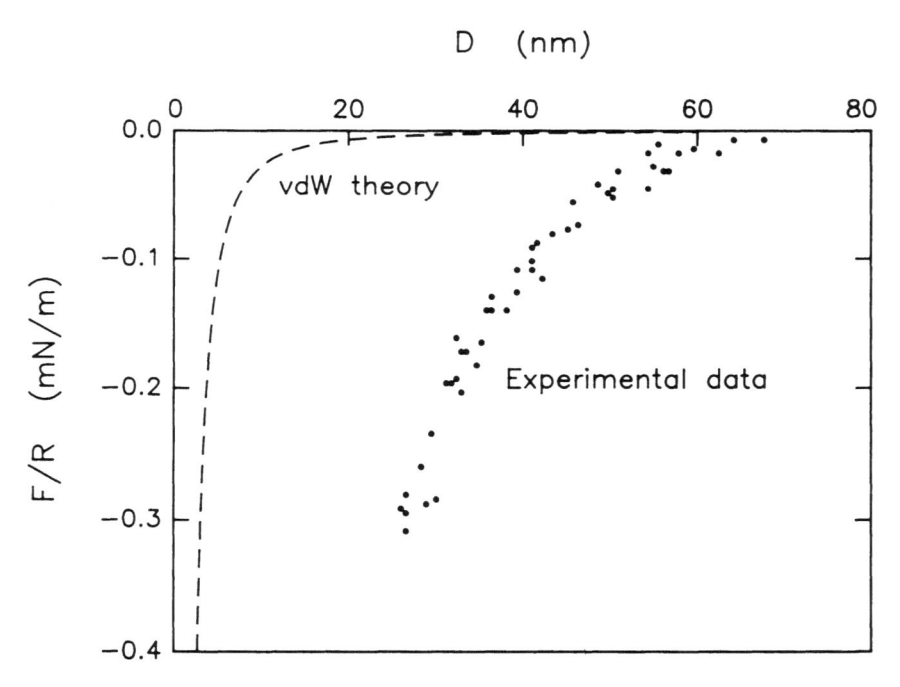

Fig. 3.9 An attractive force, much stronger than van der Waals attraction, is measured between hydrophobic surfaces immersed in water (these data are taken from [16] with permission). This so-called hydrophobic force has not yet been explained theoretically, but it is clearly an important effect to consider when dealing with hydrophobic particles.

since been confirmed by measurements on mica coated with other surfactants and on methylated surfaces of silica and mica. The force is significantly stronger than the van der Waals attraction expected for these systems, and is measurable at surface separations as large as 70 nm (Fig. 3.9) [16].

The observed attraction may well be related to the familiar hydrophobic effect in which small hydrophobic entities tend to aggregate when immersed in water. Conventionally, people have thought of that as some kind of adhesion or short-range 'contact' force; the surprising new feature of the recent measurements is that the force between hydrophobic surfaces of low curvature extends over a very large distance, so it is not clear whether the two types of hydrophobic force are related or not.

One explanation offered for the 'new' hydrophobic attraction is that it is a special kind of solvation force resulting from the particular hydrogen-bond ordering of water adjacent to a hydrophobic surface. If that is the case, then it is difficult to understand how the effect could propagate over a distance hundreds of times the diameter of a water molecule, when the hydration repulsion discussed above, also thought to be a consequence of hydrogen-bond ordering, only extends for about ten water diameters. An alternative explanation describes the long-range attraction as a novel electrostatic force resulting from correlations between patches of charge or dipoles on the surfaces, although there is some experimental evidence against this idea. A third suggestion is that the attraction is some kind of precursor to the **cavitation** of water which has been observed on separating two hydrophobic surfaces.

3.2.11 Capillary forces

It is well known that a vapour may condense in a narrow gap or capillary, and this is something that can also occur between two solid surfaces in close proximity. In the presence of a condensable vapour whose liquid wets the surfaces (by which we mean here that the contact angle θ is less than $90°$), a liquid bridge will form between the two surfaces when they are closer together than a certain distance (Fig. 3.10). The critical distance depends on the **Kelvin radius**,

$$r_k = \frac{\gamma V}{N_A kT \ln(p/p_s)} \qquad (3.47)$$

where γ is the surface tension of the liquid, V is its molar volume, and p/p_s is the relative vapour pressure of the liquid species. Note that for $p/p_s < 1$ we have $r_k < 0$, corresponding to a concave meniscus. When the liquid bridge forms, the negative curvature of its meniscus is associated with a negative Laplace pressure in the liquid

$$\Delta P = \frac{\gamma}{r_k} \qquad (3.48)$$

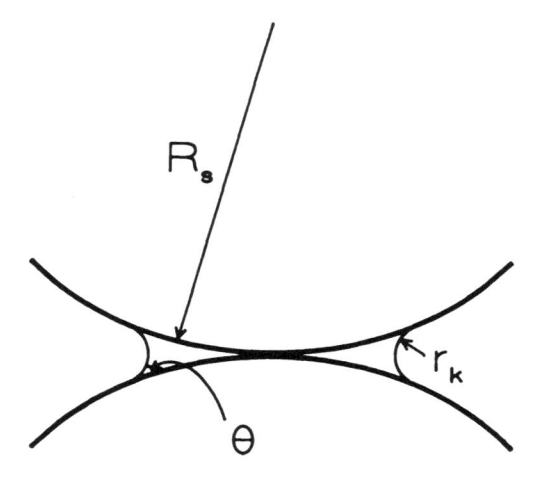

Fig. 3.10 A small capillary between two spheres, formed (for example) by condensation of water between hydrophilic particles, when the surrounding medium is moist air. Note that the same effect can also occur when the surrounding medium is not air, but a non-polar liquid. Unless the liquid has been carefully dried, it will contain enough dissolved water (a typical concentration would be $\sim 100\,\text{ppm}$) to condense in the same way. When the contact angle θ is small, the condensed meniscus results in a strong attractive force between particles.

which acts to pull the surfaces together. The net result is that with such a condensable vapour there is a strong attractive force present when the surfaces come close enough together for a meniscus to form, but the force is absent otherwise. For example, the resulting capillary force between two large contacting spheres of radius R_s with a small condensed meniscus is given approximately by

$$F = 2\pi R_s \gamma \cos \theta. \qquad (3.49)$$

When $|r_k|/R_s$ is not small the expression is more complicated, and for very large menisci the curvature could become positive and the capillary force repulsive. However, in that regime the liquid bridge would be unstable and would tend to evaporate (equation (3.47)).

The capillary force is usually larger than the other surface forces we have discussed. It is a familiar effect when handling powders in air, where the flow behaviour in particular is dramatically dependent on humidity. What is not so commonly recognized is that the same effect can occur with particles suspended in a liquid when trace amounts of a second, immiscible liquid are also present. In this case p/p_s in equation (3.47) is replaced by c/c_s, the ratio of the concentration c of the second liquid to its saturation concentration c_s, and the surface tension γ in equation (3.49) is replaced by the liquid–liquid interfacial tension.

For example, if hydrophilic particles are suspended in a non-polar oil which has been exposed to air at say 50% relative humidity, the oil will also be 50%

saturated with water, which could mean about 100 ppm of water dissolved in the oil. This water will tend to condense between particles in close proximity, i.e. phase separate from the oil, and form bridges which hold the particles together. In other words, the oil (unless it is carefully dried) can act as a reservoir of water molecules in the same way as air does, and capillary forces can be just as important in this case, exceeding any other surface forces present.

Apart from its possible important effects as a strong attractive force in particle suspensions, the capillary force plays a major role in drying and the early stages of sintering. Those are situations in which positive meniscus curvatures and repulsive capillary forces may appear, associated with thermodynamically unstable liquid bridges.

3.2.12 Effects of surfactants

The word 'surfactant' is a contraction of 'surface-active agent', a term which evokes the tendency of these molecules to adsorb to surfaces. They do this because they are amphiphilic, having a polar or hydrophilic 'head' which is soluble in water but not in oil, and a non-polar 'tail' which is hydrophobic, preferring a non-polar environment to water. When present in solution, surfactants readily adsorb to surfaces, generally with the hydrophilic end down if the surface is polar and the hydrophobic end down if it is non-polar (Fig. 3.11). This is particularly true if the surfaces have the opposite nature to the solvent, i.e. the adsorption will be particularly strong to a hydrophobic surface in water or a hydrophilic surface in a non-polar liquid.

When surfactants adsorb they 'paint' the surfaces and give them a different chemical 'colour', which can have a major effect on the forces between them. For example, hydrophobic particles in water might have a low surface charge and be pulled together by the hydrophobic attraction so that they aggregate, making it difficult to form a stable suspension. However, an appropriate amount of ionic surfactant dissolved in the water would adsorb to the particles with its hydrophobic 'tails' down, forming a monolayer and exposing charged head-groups to the aqueous phase. This would remove the hydrophobic interaction and add an electrical double-layer repulsion, thus stabilizing the suspension. The same surfactant could have the reverse effect if it were added to a stable suspension of hydrophilic particles whose charge was of the opposite polarity: it would adsorb with its ionic head-group down, neutralize the surface charge, and make the surface hydrophobic so that the particles would be likely to agglomerate. The opposite effects can be imagined for particles suspended in non-polar liquids.

These changes in surface force caused by surfactant adsorption are fairly straightforward to rationalize, though it is not always easy to predict the quantity and configuration of a given surfactant adsorbed to a given surface. For this reason it is often helpful to measure the **adsorption isotherm**, that is the amount of surfactant adsorbed as a function of solution concentration.

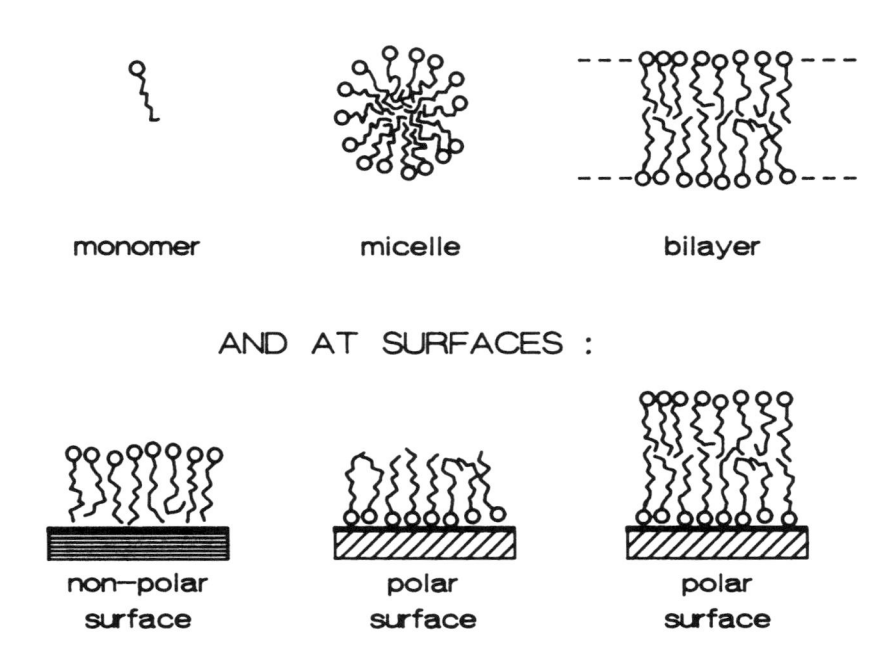

Fig. 3.11 Representation of the behaviour of amphiphilic molecules known as surfactants. One end of the molecule (drawn here as a small circle) is a polar 'head', and likes to be in or adjacent to polar media including water. The other end (shown as a zig-zag line) is a non-polar 'tail', typically a hydrocarbon chain, which prefers to be in a non-polar environment. These requirements can be met by surfactant molecules associating in solution, or by adsorbing with appropriate orientation to surfaces.

Typically there will be a certain range of concentration over which a **monolayer** of surfactant adsorbs, at a density from which it can be construed that the elongated molecules are oriented perpendicular to the surface. This is the coverage required for the effects described above. At lower concentrations there will be submonolayer coverage, with the possibility of having surfactant molecules lying parallel to the surface. At higher concentrations it is possible to get a second monolayer adsorbed on the first, generally in the opposite orientation. For example if the first monolayer forms on a polar surface with its head-groups down and hydrophobic tails out, rendering the particle hydrophobic, a second monolayer would form with its tails down and head-groups out, making the surface hydrophilic again.

Surfactant adsorption can affect other forces between surfaces, in particular the van der Waals attraction. If, as is often the case, the refractive index of the surfactant is intermediate between that of the particle and that of the solvent, an adsorbed layer of surfactant will weaken the attraction. Furthermore, the finite thickness of the adsorbed monolayers prevents the particle surfaces from coming into close contact where the van der Waals attraction is strongest, an important effect called **steric stabilization**. Keeping the particles slightly separated may be enough to prevent them from adhering. Unless it is done very carefully, however, such fine tuning of the van der Waals interaction may be completely overwhelmed by the more dramatic changes in surface forces resulting from monolayer adsorption, as described above.

With all of these possibilities, the addition of surfactants to the suspending medium is one of the most powerful methods of modifying surface forces. It is not easy in every case to control the forces in exactly the way that one would like, because of all the different phenomena that might occur in the same system. Nevertheless, much can be done if the various effects are properly understood, and surfactants have become popular as dispersing agents.

3.2.13 Effects of polymers

The presence of polymers can affect surface forces in varied and subtle ways. Theories are available to describe many of the phenomena [17], but they are generally too complicated to be reproduced here, so we will attempt only a qualitative description of the effects of polymers.

In considering forces between surfaces separated by a polymer **solution**, it is important to know whether or not polymer molecules adsorb to the surfaces, which depends on whether segments of the polymer molecule prefer to be in contact with the surface or with the solvent. Adsorption is likely to occur when the solvent is 'poor', i.e. the polymer solubility is low. On the other hand, polymer is less likely to adsorb from a 'good' solvent which has a high affinity for the polymer segments.

The simplest effect of **adsorbing** polymers is to coat particles with a polymer layer, giving rise to **steric stabilization** (Fig. 3.12). If enough polymer adsorbs, the thickness of the coating is sufficient to keep particles separated by steric repulsions between the polymer layers, and at those separations the van der Waals forces are too weak to cause the particles to adhere. For small particles, Brownian motion is then sufficient to keep them suspended indefinitely.

Steric stabilization requires adsorption of a reasonably dense polymer layer on each particle, which in turn requires a sufficient supply of polymer in the solution. Good adsorption can be achieved even with weak attractions of each segment to the surface because there are many segments available to make the overall attraction of the **molecule** quite strong. Very strong segmental adsorption can in fact be deleterious, because if polymer segments stick strongly and irreversibly to the first area of surface that they encounter, further polymer is prevented from adsorbing in that vicinity, and a comparatively poor coverage can result. Weaker adsorption allows some mobility on the surface or rearrangement of adsorbed segments to accommodate further adsorption until a dense layer is formed.

There is another possible effect which results from strong adsorption at low polymer concentration, particularly with high molecular weight polymer. This

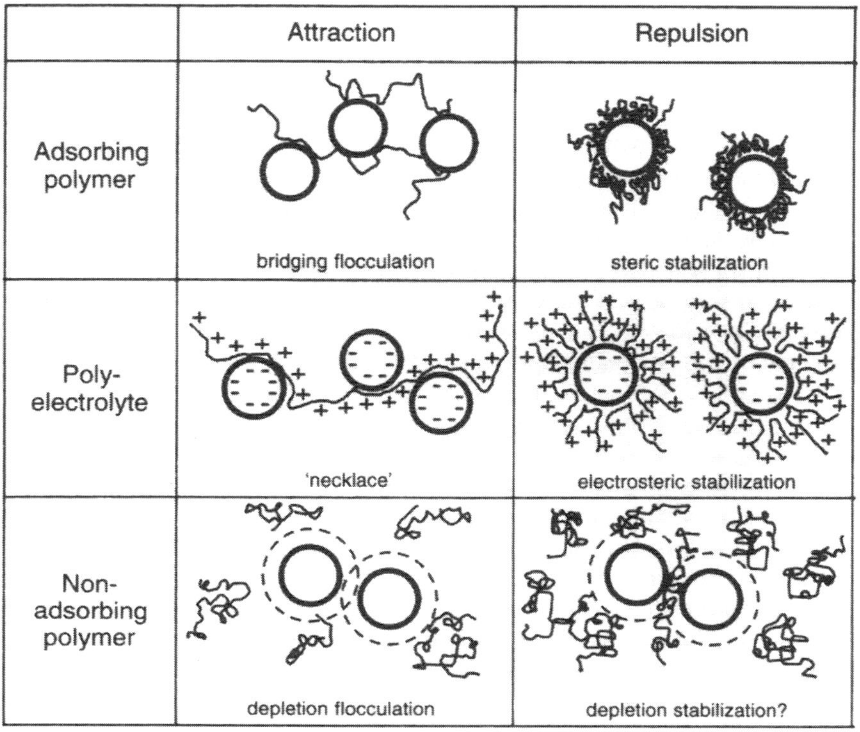

	Attraction	Repulsion
Adsorbing polymer	bridging flocculation	steric stabilization
Poly-electrolyte	'necklace'	electrosteric stabilization
Non-adsorbing polymer	depletion flocculation	depletion stabilization?

Fig. 3.12 Various interparticle forces are possible when polymer molecules are present, depending on the polymer concentration and molecular weight, whether or not the polymer adsorbs to the particles, and whether the polymer is charged. Low concentration and/or high molecular weight tend to give attractive bridging (or necklacing) leading to aggregates of particles, often undesirable for ceramic processing requirements. The effective attraction giving rise to depletion flocculation comes from a reduced osmotic pressure in the region between particles when non-adsorbing polymer is excluded from that region. The possibility of depletion stabilisation, arising from the entropic penalty of compressing non-adsorbing polymer between particles at high polymer concentration, is still open to question.

effect is the possibility that one polymer molecule can adsorb to more than one particle at the same time, thereby forming a link which holds the particles together. This important effect, known as **bridging flocculation**, destabilizes a suspension (Fig. 3.12) and so is generally undesirable for ceramic processing.

The use of **block copolymers** can overcome this problem. These molecules consist of two different polymer chains grafted together end-to-end, so that they bear a similarity to the amphiphilic surfactants described in the previous section. If a block copolymer is dissolved in a liquid which is a good solvent for one of its ends but a poor solvent for the other, the latter end will have a strong tendency to adsorb to particles, while the remainder of the molecule extends into the solvent. Thus a coat of non-adsorbing and non-bridging polymer is attached to the particle, effectively preventing other particles from approaching. (Unfortunately, block copolymers tend to be more difficult to synthesize, and thus more expensive and less readily available than homopolymers.)

Still more complicated effects can arise if polyelectrolytes are used. The addition of large polyions to a suspension of charged particles of opposite sign is likely to lead to bridging with one polymer chain attached to two or more particles (forming a 'necklace'), and hence flocculation. On the other hand, adsorption to neutral particles will give those particles a large charge and thus an effective electrostatic stabilization mechanism to add to the steric stabilization: a combination known as **electrosteric stabilization**. The charge will be distributed along the adsorbed polymer chains; at low electrolyte concentration the chains repel each other and also other parts of themselves, causing polymer 'hairs' attached to particles to stand on end. At high electrolyte concentration the hairs may collapse back towards the particle surface, thus diminishing the particle's effective diameter at the same time as the range of the double-layer repulsion between particles is reduced.

Let us now go back to considering simple polymers, this time dissolved in a good solvent so that they are **non-adsorbing**. Because of restrictions on their possible configurations, polymer molecules tend to stay out of a region near the surfaces, known as the depletion layer. As two particles approach, polymers are evicted from the gap between their surfaces, decreasing the concentration in the gap and increasing it elsewhere. This creates an osmotic pressure difference which tends to push the particles together, i.e. it creates an effective attractive force leading to **depletion flocculation**. It has been argued [17] that at higher polymer concentrations there would be a second effect arising from the entropic penalty of compressing the free polymer chains between particles, if not all of the chains are excluded from the gap. This would give a repulsive force and lead to **depletion stabilization**. However, this idea is not universally accepted, and it remains an open question whether this effect occurs in practice.

Even without the last possibility, it is clear that the addition of polymers to a suspension can have various effects on surface forces, depending on the chemical nature of the polymer, the solution and the surfaces, as well as on molecular weight, temperature, and particle concentration. For this reason the use of polymers is another powerful and popular technique for controlling the stability of suspensions. The theoretical principles are reasonably well understood, although the detailed calculations are often very difficult and it is not always possible to predict which type of force will dominate under a given set of circumstances.

3.2.14 Short-range forces

When atoms, molecules or surfaces come very close together there are strong forces between them as a result of the overlap of their electron clouds. This may result in chemical bonding or hydrogen bonding, which in the language of this article are equivalent to strong, very short-range attractive forces; there will also be Born repulsion, which is a very short-range, strong repulsive force like a hard wall. Neither of these contributions to the 'surface force' has a range of more than 0.1 or 0.2 nm, so we can safely consider them as 'contact' forces.

The forces enumerated in previous sections all act over a longer range, and we have paid attention to them because it is those forces which determine the stability of suspensions. The contact forces are important in determining what happens **after** particles have agglomerated, as well as in other key areas such as adhesion, fracture and friction. Because contact forces depend very much on the specific chemistry of the molecules or surfaces involved, it is impractical to offer general theories to describe their magnitude or even their sign.

It should be noted that the short-range forces also have important **indirect** effects on some of the longer-range forces we have discussed. For example, hydrogen bonding or chemical bonding between solvent molecules and a surface will determine the nature of solvation forces. Bonding between surfactant or polymer molecules and a surface will be a prime factor in determining the adsorption of those molecules, which as we have seen has a dramatic effect on the force between two such surfaces. Short-range forces also determine whether molecules adsorb from a vapor, what the contact angle of a liquid condensed on a surface is, and hence what the capillary force will be. Born repulsions manifest themselves in finite-size effects which are the basis of steric repulsions between adsorbed surfactant or polymer layers, and structural forces.

3.2.15   Present status of theoretical and experimental work on surface forces

In each of the sections above we have tried to give some indication of the status of our theoretical understanding of the different contributions to forces between particles in suspension. Our knowledge of van der Waals forces, electrical double-layer forces, capillary forces and most of the polymer effects is very sound. However, our ability to **predict** the magnitude of these forces in a given system depends on the availability of certain data, which are not always easy to obtain. For example, to compute the van der Waals force we would need very detailed optical data for all the materials involved; for the double-layer force we would need to know the surface charging behaviour and electrolyte strength; for capillary forces we need the contact angle and the water content in the solvent; and for polymer interactions we need details such as amount adsorbed and polymer segment density distribution in the adsorbed layer.

There is not really any theory for surfactant effects, although we can have an educated guess as to what those effects will be if we know how much surfactant adsorbs to the particles. Solvation forces cannot in general be predicted from any existing theory, although again it may be possible to guess what they will be, qualitatively at least, from a knowledge of the nature of the solid surfaces and the liquids involved. Hydrophobic forces remain mysterious, but once again we could expect that they will be important whenever hydrophobic particles are suspended in an aqueous liquid.

On the experimental front, many indirect inferences about surface forces in particular systems have been drawn from measurements of the stability, turbidity and rheology of suspensions, as well as from the swelling behaviour of clays and of surfactant phases. There are also methods of measuring surface forces directly in model systems. The most popular of these methods employs a device designed by Israelachvili and Adams [11], in which two molecularly-smooth sheets of mica are mounted as crossed cylinders, and brought into close proximity in a liquid or vapour environment. The separation between the mica surfaces is measured with an accuracy of ~ 0.1 nm using a sensitive optical interference method, and surface forces are measured to 100 nN by monitoring the deflection of a cantilever beam on which one of the micas is mounted.

This apparatus has been used since 1978 to measure all of the effects described in the preceding sections, and the results have added significantly to our knowledge of surface forces. Some of the measurements have helped substantiate the existing theories (Fig. 3.8); others have thrown up novel results which have challenged the theoreticians, particularly on structural, solvation [12] and hydrophobic forces [16] (Fig. 3.9), and on surfactant effects. The same technique has recently been extended to sapphire [13] and silica [14] surfaces.

3.3 BEHAVIOUR OF SUSPENSIONS

3.3.1 Dispersion

In the colloidal processing of ceramics, particles must first be dispersed in a liquid. This enables homogenization, settling of large undispersed aggregates or other impurity particles, and mixing of different particles if required. To do this we need to arrange solution conditions in the liquid so that the suspension will be stable, i.e. we need to arrange one or more mechanisms of repulsive force to overcome the van der Waals attraction, while avoiding any other attractions such as polymer bridging, capillary forces or hydrophobic effects. Dispersion usually also requires an input of mechanical energy to break up existing aggregates.

3.3.2 Dense suspensions

In ceramic processing it is usually desirable to form rather dense suspensions, or slurries, meaning that there are many particles in a limited volume of liquid. For this reason it is pertinent to make a few remarks here about dense suspensions.

In addition to understanding the pairwise interactions between two particles, it becomes necessary to consider the statistical mechanics of a collection of those particles in a given volume. This introduces such concepts as many-body interactions (e.g. how the force between two particles is affected by a third particle nearby) which are much harder to compute than the two-body interactions we have been discussing. However, while quantitative calculations become very difficult, the qualitative nature of the forces such as their sign, range and relative magnitude, can still be applied to dense suspensions.

The results of statistical mechanics arguments are presented in terms of **phase** behaviour of the suspension: whether there is one phase (the whole suspension having the same microstructure) or two (for example, separated into a sediment and a supernatant liquid), and what the phases are. A supernatant with a low density of particles would be thought of as a 'vapour', while a dense phase could be 'liquid' when the particle packing is disordered, or 'solid' if the particles are packed into an ordered array. There can also be non-equilibrium phases, for example, a volume-filling **gel** in which particles are connected together by attractive forces, but held in an open network structure like a scaffold.

There is one way to think of particles in suspension which may be useful for understanding (at least qualitatively) the structure and later the rheology of slurries. The particles themselves have a certain size, which we have no difficulty in comprehending, and which we will call the **hard size**. One particle cannot enter the volume occupied by another: the size is the 'hard wall' diameter (or other dimension, such as length, for non-spherical particles). We have also learnt that particles may interact through one or more surface forces, so that they may begin to 'feel' each other's presence at a longer distance which depends on the range of the force. Thus each particle may be thought of as having an **interaction size** which is something like its diameter plus twice the range of the surface force (Fig. 3.13). The space occupied by each particle could be characterized both by its 'hard volume' and its 'interaction volume'.

A suspension containing N particles could only be single-phase if its total volume is less than approximately N times the interaction volume. If the interaction volume is significantly larger than the hard volume, as with long-range double-layer forces, for example, it will probably be fairly easy to compress the single-phase slurry since the particle interactions are rather 'soft' and the particles will not be densely packed.

On a more pragmatic level, there are two other points to be made about dense suspensions. The first is that while there are methods to characterize particle properties of significance for particle interactions (e.g. electrophoretic mobility to measure surface potential, light scattering to determine particle size) which can be used effectively in dilute suspensions, these methods do not work well in concentrated suspensions. The second is to note that the properties of the suspending liquid can be markedly influenced by the presence of a large number of particles. For example, adsorption or desorption of ions at the particle surfaces can significantly alter the electrolyte strength or pH of

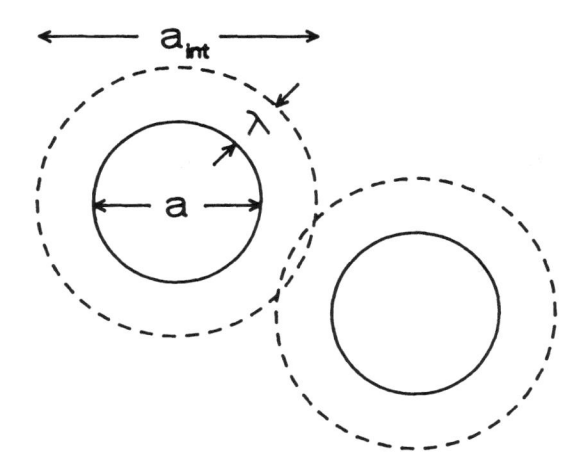

Fig. 3.13 To understand the rheology of a suspension, it is helpful to think of particles as having an **interaction size**, a_{int}, larger than their **hard size**, a. Factors such as the solids loading (volume fraction) of the suspension are calculated from the hard size. However, the structure and rheology of the suspension are determined by the interactions between particles, and therefore depend on the interaction size, which includes the range, λ, of the interparticle forces.

an aqueous solution. What determines surface forces and suspension stability is the solution properties **after** the particles have been added, which may be quite different from the solution that was initially prepared. Conversely, a large volume fraction of small particles has a high surface area, which can 'mop up' a large amount of adsorbate such as surfactant or polymer. If these are required as stabilizing agents, one must be sure to add enough to coat all of the particles, because the solution will be significantly depleted by the adsorption.

3.3.3 Consolidation

Once a homogeneous slurry has been formed, it must be consolidated to a high solids loading before final densification to form the ceramic. Consolidation requires increasing the volume fraction of particles, and this can be achieved in different ways.

One way is to change the solution conditions of the suspending liquid so that the interparticle forces become **attractive**. In many cases an appropriate agent (e.g. a polymer) is added to cause the particles to flocculate or stick together. Note that removing any repulsions (e.g. eliminating double-layer forces by changing pH to the IEP) will achieve this, since the van der Waals attraction is always present. One might at first think that **strong** attractions would maximize the volume fraction, but this is not the case. In fact if the attractive forces are too strong, colliding particles stick rigidly where they first make contact. The next particle that comes along is unlikely to arrive and stick at exactly the position required for a close-packed configuration, and the same is true for all subsequent particles. The result is an agglomerate which has an open structure and a low density, sometimes called a **floc**. Denser packing are achieved if the particles can slide around each other after they are in contact so that the aggregate can rearrange itself; a **weak** attractive force is desirable to achieve this. If some liquid, or an adsorbed layer with a low shear strength, remains on the particles when they are agglomerated, the sliding required for such rearrangement will be greatly facilitated.

For this reason it is not desirable to have aggregation into a strong primary minimum (Fig. 3.7) in which the particle surfaces would come into close contact due to van der Waals or hydrophobic attraction. It would be preferable to have particles in a secondary minimum, in which a short-range double-layer repulsion is still present to keep the particle surfaces separated by a few nanometres; or a 'solvation minimum' which is similar except that the short-range repulsion comes from a solvation force.

A second way to increase volume fraction is to remove some of the liquid, for example by evaporation, by slip casting or by pressure filtration. This has the effect of compacting the particles even if there is no attraction between them. In fact, high volume fractions can be attained when the interparticle interactions are purely repulsive. Repulsive forces imply the retention of some stabilizing medium between the particles (the difference between the 'interaction size' and the 'hard size'). This could be electrolyte solution, adsorbed surfactant or polymer, or some liquid of solvation, and it will become progressively more difficult to remove liquid from the consolidated slurry as the particles are forced closer together. However, this situation has the benefit that the liquid remaining between the particles can lubricate the particle 'contacts', enabling particle rearrangement. Furthermore, a homogeneous microstructure will be maintained as each particle tries to remain as far away as possible from all of its neighbours. The remaining liquid will also help to bind the consolidated body together (section 3.3.5).

3.3.4 Rheology

The flow properties of suspensions, which are extremely important for processing and forming considerations, are determined by the volume fraction, size distribution and interparticle forces. For dilute suspensions there is a fairly straightforward relationship between volume fraction and viscosity, but for dense suspensions there are no simple formulae for predicting the rheological properties [4]. Some of the trends can be understood as follows, using the concepts of hard size and interaction size introduced in section 3.3.2 (Fig. 3.13). Of course it is the hard volume fraction that is usually calculated as the volume fraction, and which is the important number in determining solids loading of a slurry.

The viscosity of a suspension depends on collisions between particles. Since a collision does not necessarily involve hard-wall contact between particles, but could be mediated by **any** interparticle force, it is the interaction volume fraction which is important in determining the viscosity at low shear rates.

At higher shear rates the number of collisions between particles increases. The suspending medium applies hydrodynamic forces to the particles, and as shear rate increases these forces become more significant, eventually dominating over the interparticle forces. One could think of the interaction size shrinking down towards the particle size as shear rate increases; thus the limiting high-shear rate viscosity depends on the hard volume fraction.

Hydrodynamic shearing forces can also overcome attractive interactions, and break up flocs or agglomerates of particles. When a concentrated slurry has a network of interacting particles filling the volume (e.g. a single solid-like phase, or a gel) it becomes capable of supporting shear, and has a non-zero **yield stress**. Typically in this kind of system the network can be broken up by the application of shear. The shear stress required to do this depends on the strength of the attractive forces.

Once the yield stress has been exceeded, the particle network may take some time to re-form. If shear is maintained the suspension will flow more easily than it did initially: we say the suspension is **shear-thinning**. This is often a desirable property, since a slurry with sufficient yield stress can be handled as a solid body, but it can also be made to flow when sheared, which facilitates transport and forming processes.

3.3.5 Strength of consolidated body

The consolidated material should retain sufficient strength and deformability so that it can be handled and, in many cases, formed to shape by a process such as moulding. An agent that helps to hold the consolidated body (or 'compact') together is called a **binder**. In most cases, binders act by exploiting a negative capillary pressure, by polymer bridging between particles, or possibly both.

3.4 CONTROLLING INTERPARTICLE FORCES

Having discussed the effects that different surface forces can have on suspensions, we now want to consider methods of controlling those forces and hence the suspension properties. As we have seen, both for dispersing particles and for subsequently consolidating them, it is generally desirable to minimize the attractive forces. Since van der Waals forces are always present, this means that we must exploit some kind of repulsive force to counterbalance or to overcome the van der Waals attraction, and we should be wary of introducing other attractions. In adding repulsive forces, other considerations come into play, the suspension rheology in particular. In fact most of the methods of minimizing attractions and maximizing repulsions have already been alluded to in the foregoing discussions of each type of force, so this section really only reiterates what has already been said.

3.4.1 Minimizing attractive forces

There is only one true way to reduce the van der Waals force, and that is to choose a suspending medium with dielectric properties (particularly at optical and UV frequencies) as close as possible to those of the solid. Of course there are serious limits to how close a match one can obtain, particularly since not all liquids will be suitable as suspension media. In many cases, perhaps even most cases, questions of chemical compatibility, solvation properties, solvent quality, cost, safety, etc. will make this approach unpractical.

The alternative is to keep particles separated to a distance at which the van der Waals force is weak, for example by steric stabilization with a surfactant layer, or by exploiting a solvation force. This approach, which is far more common and flexible, really falls into the category of adding a repulsive interaction (section 3.4.2).

There are other attractive forces which might occur under specific circumstances, and so one should remain aware of them and ensure that they are not inadvertently introduced. First is the hydrophobic interaction, which will occur between hydrophobic particles in an aqueous suspension. Adding a surfactant to a suspension of hydrophilic particles in order to reduce van der Waals attraction could well coat the particles with a monolayer and render them hydrophobic, which would have the undesired effect of aggregating them.

Second is the capillary force, which can occur with hydrophilic particles when there are trace amounts of water in a non-polar suspending liquid. To avoid this, it may be necessary to use a desiccant to remove water from the liquid, and to avoid exposure to further sources of water (including air).

Third come the possibilities of polymer bridging and depletion flocculation. Polymers can be very effective and useful stabilizers, so long as each particle has a reasonably dense coating of the adsorbing polymer. However, the same polymer under more dilute conditions can attach to more than one particle and cause flocculation. Whether or not this occurs can depend on how the suspension is formed: adding a small amount of adsorbing polymer to a dense suspension would be likely to destabilize rather than stabilize it; adding particles progressively to a polymer solution would effectively coat each particle before it had an opportunity to stick to its neighbours. Non-adsorbing polymers can destabilize a suspension by depletion flocculation.

Fourth (and probably least important) is the ion-correlation effect described in section 3.2.7. For highly-charged surfaces, the double-layer repulsion would be less than expected when there are polyvalent counterions in the solution.

3.4.2 Arranging repulsive forces

Particles suspended in a polar liquid (usually water) generally experience a double-layer repulsion, and this repulsion can be controlled by changing the solution conditions. For most ceramic materials the surface charge will depend on pH, typically being positive in acidic conditions and negative when the solution is alkaline. At some point in between (the IEP), the charge will be zero (Fig. 3.5) and the double-layer repulsion will vanish, which provides a common method for coagulating suspensions. Stability is maximized by operating at a pH far from the IEP.

The range of the double-layer repulsion is reduced by increasing electrolyte concentration (do not forget the H^+ and OH^- ions: the Debye length is short at extreme values of pH). In concentrated electrolyte the range can be short enough to allow a secondary minimum in the force at a finite surface separation (Fig. 3.7). This can be very convenient because it allows reversible coagulation, and because it allows compaction to a high volume fraction by avoiding the open floc structures associated with attraction into a primary minimum (section 3.3.3). However, while the double-layer repulsion is very flexible and easy to modify, there are situations (e.g. large or dense particles, non-polar solvents, low surface charge) in which it is simply not strong enough to stabilize a suspension.

Addition of suitable surfactants is another powerful and flexible method for creating a stable suspension. The immediate effect that surfactants have is to form a layer on the particles and thereby prevent them from coming close enough together for the van der Waals force to be strong. Ionic surfactants can also increase surface charge and so enhance electrostatic stabilization. In many cases the real benefit of surfactants is in reversing the 'chemical polarity' of the surface, for example in coating hydrophobic particles and rendering them hydrophilic so that they can be stabilized in water.

To be effective, the surfactant must adsorb readily to the solid particles. This usually is the case, but anionic surfactants, for example, would not adsorb to a negatively-charged surface unless some divalent cation were present to provide ionic bridging. In using surfactants, care must be taken to avoid inappropriate changes in the hydrophilic/hydrophobic nature of the surface, which could easily destabilize rather than stabilize the suspension.

Polymers also provide a rich variety of agents for stabilization, through steric effects with adsorbing polymers and depletion stabilization with non-adsorbing polymers. The first characteristic to be considered in selecting a polymer is whether or not it will adsorb to the particle, which depends largely on whether the liquid is a good solvent or a poor solvent for that polymer. The range of interaction increases with the polymer molecular weight. As discussed above, attention needs to be given to the possibility that attractions can also be introduced by polymers, which is undesirable for dispersion but may be useful for consolidation (whereupon the polymer may act as a binder).

One very effective repulsion comes from solvation of particles. Solvation forces are strong and short-range repulsions, which is an ideal combination for producing stable, dense suspensions with minimal viscosity for a given (hard) volume fraction. Unfortunately, solvation effects are not always under our control, nor are they easily predictable. However, a few general comments can be made.

It has been noted above that it is difficult to disperse hydrophobic particles in aqueous liquids, and it can be difficult to disperse hydrophilic particles in non-polar liquids (because double-layer forces are less effective, and capillary forces can be a problem). As a general rule, it will be far easier to disperse particles whose surfaces are alike in chemical nature to the suspending liquid: polar surfaces (e.g. oxides) in polar liquids, non-polar (e.g. hydrocarbon) surfaces in non-polar liquids. The chemical similarity can be judged by whether the liquid **wets** the surface. The better the wetting, the lower the solid–liquid interfacial energy, which means less adhesion between particles immersed in that liquid. If the liquid wets the particle completely (zero contact angle and a thick wetting film) the interfacial energy may in fact be **negative**, which implies the existence of a repulsive solvation force. In that case the particles would disperse **spontaneously**.

3.4.3   Other considerations

In the discussion so far we have alluded several times to secondary effects or side effects of the methods that could be used to modify surface forces. The first and probably the most important is unwanted side effects of introducing surfactants or polymers, when an attraction could be introduced instead of the desired repulsion.

A second consideration is the possible symbiotic or antisymbiotic relationships that may exist between different additives. We have not discussed this so far because it is a complex and sometimes unpredictable issue. An example would be that the amount of electrolyte in solution could affect the aggregation properties and consequently the adsorption of a surfactant, or it could alter the conformation of a polyelectrolyte.

In introducing repulsive forces into a particle suspension, one would want to consider how the **range** of the repulsion affects the suspension properties, in particular its solids loading and its rheology. For a given volume fraction of particles, a long-range repulsion would lead to a higher viscosity than would a short-range repulsion, and could make it difficult to increase the loading to an acceptable level for processing.

Finally, in a related issue, one must consider what happens to the suspending liquid, including stabilizing additives, when the ceramic is to be densified. It is the solids which provide the useful material in the slurry; most if not all of the liquid and the additives must eventually be removed during drying and firing. Clearly, the less liquid that remains, the less difficult this is going to be. For this reason, it is desirable to minimize the difference between the interaction volume and the hard volume of the particles. Another consideration is the chemical nature of the stabilizing additives. Those additives that do not evaporate during drying will remain on the particle surfaces, and many will eventually be burnt. Will their volume be so great that they cause problems during burnout? Will the combustion products be detrimental to the final ceramic product? Or could stabilizing agents be chosen that are actually a beneficial component or an integral part of the ceramic material?

3.5 SOME EXAMPLES

In the final section of this chapter, a few examples selected from the recent ceramics literature are used as illustrations of cases in which interparticle interactions have been modified in order to produce desirable features in the processing or properties of ceramic materials. Note that the choice of examples is arbitrary and is not intended to be complete; other equally interesting examples have been omitted.

3.5.1 Improved strength from homogenization of powder

It is well known, following the arguments of Griffith, that the strength of a brittle material is limited by the size of the largest flaw present. A common source of flaws in ceramic materials is the presence of agglomerated particles prior to sintering. A recent demonstration of this well-known principle has been presented by Alford, Birchall and Kendall [18], who measured the bend strength of alumina prepared from aqueous dispersions stabilized by poly(vinylalcohol acetate). Different sources of alumina powder and different extrusion procedures led to different sizes for the largest flaw observed, and the measured bend strength of the finished bar increased from 0.3 to 1.0 GPa as the maximum flaw size decreased from 100 to 15 μm.

3.5.2 Aggregation with strong attractive forces gives low volume fraction

Philipse, Bonekamp and Veringa [19] have given an illustration of the fact that particles aggregated by strong attractive forces form an open gel structure with high porosity and low density. They studied suspensions of two types of solid, silica spheres and alumina particles, each prepared in two ways to give either dispersed or aggregated suspensions. Under the influence of long-range double-layer repulsions, silica particles form an ordered, volume-filling array (a 'colloidal crystal'); this is easily compacted to high density (random close-packed spheres) by colloidal filtration to remove as much solvent as possible. When electrolyte is added to reduce the double-layer repulsion, the silica spheres aggregate, but probably only in a weak secondary or solvation minimum, and filtration again gives high solid density. Alumina dispersed in water by a polyelectrolyte does not form a colloidal crystal because the particles are more polydisperse and irregular in shape, but nevertheless this suspension can be compacted under filtration. However, without the polyelectrolyte present, alumina particles aggregate into a primary minimum and form a volume-filling gel. The volume fraction (solid loading) can be increased by colloidal filtration, but not to the same value, and the compact remains porous.

3.5.3   Adding the right amount of polymer dispersant

Novich and Pyatt [20] investigated alumina particles dispersed in an aqueous solvent by adding a commercial polyanionic dispersant. Measuring the viscosity as a means of monitoring the degree of dispersion, and presuming that for a given volume fraction of solids lower viscosity reflects better dispersion, these authors show that there is an optimum concentration of polyelectrolyte (0.75wt% of the suspending fluid). At lower concentrations there is probably polymer bridging of particles that causes some degree of agglomeration, while at higher concentrations the thickness of the adsorbed polymer layer and the excess polymer in solution both act to increase viscosity.

3.5.4   Effective use of block copolymer

A clear illustration of the benefits of using a block copolymer is given by Kerkar, Henderson and Feke [21]. Endeavouring to disperse silicon in non-polar solvents (benzene or trichloroethylene), these authors found that polystyrene (PS) (for which these are good solvents) does not adsorb strongly to the particles and is not effective as a dispersant. With poly(methyl methacrylate) (PMMA), on the other hand, these are poor solvents; the polymer does adsorb strongly to the particles. PMMA gives effective dispersion, but not at low concentration where interparticle bridging may be a problem. The most effective solution is to use a block copolymer of PS–PMMA, the PMMA end of which adsorbs strongly while the PS end extends out into solution, providing effective dispersion even at low concentrations, with no possibility of bridging.

3.5.5   Controlling rheology as well as stability

Velamakanni *et al.* [22] make a distinction between aggregated and coagulated systems, meaning that in the former, particles are held together by strong attractive forces, while in the latter only weak attractions are acting. In an aqueous suspension of alumina particles, they present evidence that a short-range hydration repulsion (section 3.2.9) acts at low pH to prevent particles from coming into a primary minimum, instead remaining in a much weaker 'hydration minimum'. In this state, some water remains between the particles, and the repulsive force provides a degree of lubrication, allowing the particles to move around each other and to be compacted to a high volume fraction (comparable to a dispersed system, higher than an aggregated one) by pressure filtration. At the same time, the particles are close enough for van der Waals forces to be significant, linking them in an attractive network and giving rise to a high viscosity (and probably a finite yield stress). The authors suggest that this is a useful feature because it prevents mass segregation by sedimentation. One suggested application is to the processing of composites, because other, denser, particles of another phase (including fibres or whiskers) could be suspended uniformly in the slurry.

3.5.6 Maintaining low viscosity until after forming

Young *et al.* [23] (Chapter 6) present a novel and very promising new idea for incorporating a binder into a dispersion without sacrificing the low viscosity desirable for moulding at high solids loading. Polymerizable organic monomers (acrylamides) are added to an aqueous dispersion of alumina stabilized by polyelectrolyte. Since the monomers have low molecular weight, they have only a slight effect on the viscosity of the dispersion, and the slurry can be cast easily at high solids loading. After casting, polymerization is initiated, and the resultant polymer holds the ceramic particles together in strong compact (green body). This process reduces the mechanical energy required for moulding, it requires a comparatively small amount of binder, and it allows near-net-shape forming of complex shapes.

3.5.7 Using 'one stone to kill two birds'

A very interesting effect, in which one processing additive serves two purposes, has been described by Lidén *et al.* [24]. With the aim of adding a small amount of yttria to silicon nitride to act as a sintering agent, these authors note that the conventional method of co-milling yttria and silicon nitride particles gives poor homogeneity of yttria in the sample, because the yttria and silicon nitride particles are of comparable size. A better alternative is to use much smaller yttria particles (10 nm) prepared by a sol technique. The yttria particles are charged positively in water at pH 7, whereas silicon nitride is negatively charged. Thus, adding the yttria slowly to a silicon nitride dispersion, yttria particles adsorb to the much larger silicon nitride particles, effectively dispersing the yttria homogeneously and placing it at the particle interfaces where it is required. Above a certain concentration (3wt% yttria), the overall sign of the composite particles changes sign. Contrary to expectations, a small amount (1%) of yttria **increases** the degree of dispersion even though it reduces the effective surface potential of the larger silicon nitride particles, and indeed the dispersion is still stable when the particles are neutral. The reasons for the increased stability of such a heterodispersion are not clear; nevertheless, it turns out that yttria acts both as a dispersant and as a sintering agent.

While it appears that the discovery of these dual benefits may have been serendipitous in this case, it does provide a suitably optimistic note on which to end this chapter. Given an appropriate understanding of interparticle forces, coupled with a lively imagination, one might well be able to find other examples in which one ingredient can serve more than one purpose, or where two ingredients may act symbiotically so that their combined benefit is greater than the sum of their individual effects. In such a case, perhaps it would be appropriate to refer to the ingredients not as 'additives', but as 'multipliers'.

REFERENCES

1. Lange, F.F. (1989) Powder processing science and technology for increased reliability. *J. Am. Ceram. Soc.*, **72**, 3–15.
2. Israelachvili, J.N. (1991) *Intermolecular and Surface Forces*, 2nd edn, Academic Press, London.
3. Hunter, R.J. (1987) *Foundations of Colloid Science*, Vol. 1, Oxford University Press, Oxford.
4. Russel, W.B., Saville, D.A. and Schowalter, W.R. (1989) *Colloidal Dispersions*, Cambridge University Press, Cambridge.
5. Reed, J.S. (1988) *Introduction to the Principles of Ceramic Processing*, Wiley Interscience, New York.
6. Horn, R.G. (1990) Surface forces and their action in ceramic materials. *J. Am. Ceram. Soc.* **73**, 1117–35.
7. Mahanty, J. and Ninham, B.W. (1976) *Dispersion Forces*, Academic Press, London.
8. Parsegian, V.A. (1975) Long-range van der Waals forces, in *Physical Chemistry: Enriching Topics from Colloid and Surface Science* (eds H. van Olphen and K.J. Mysels), Theorex, La Jolla, California, pp. 27–72.
9. Hough, D.B. and White, L.R. (1980) The calculation of Hamaker constants from Lifshitz theory with applications to wetting phenomena. *Adv. Colloid Interface Sci.*, **14**, 3–41.
10. See for example Chan, D.Y.C., Pashley, R.M. and White, L.R. (1980) A simple algorithm for the calculation of the electrostatic repulsion between identical charged surfaces in electrolyte. *J. Colloid Interface Sci.*, **77**, 283–5.
11. Israelachvili, J.N. and Adams, G.E. (1978) Measurement of forces between two mica surfaces in aqueous electrolyte solutions in the range 0–100 nm. *J. Chem. Soc. Faraday Trans. 1*, **74**, 975–1001.
12. Pashley, R.M. (1981) DLVO and hydration forces between mica surfaces in Li^+, Na^+, K^+ and Cs^+ electrolyte solutions: a correlation of double-layer and hydration forces with surface cation exchange properties. *J. Colloid Interface Sci.*, **83**, 531–546.
13. Horn, R.G., Clarke, D.R. and Clarkson, M.T. (1988) Direct measurement of surface forces between sapphire crystals in aqueous solutions. *J. Mater. Res.*, **3**, 413–16.
14. Horn, R.G., Smith, D.T. and Haller, W. (1989) Surface forces and viscosity of water measured between silica sheets. *Chem. Phys. Lett.*, **162**, 404–8.
15. LeNeveu, D.M., Rand, R.P. and Parsegian V.A. (1976) Measurement of forces between lecithin bilayers. *Nature*, **259**, 601–3.
16. Christenson, H.K. and Claesson, P.M. (1988) Cavitation and the interaction between macroscopic hydrophobic surfaces. *Science.* **239**, 390–92.
17. Napper, D.H. (1983) *Polymeric Stabilization of Colloidal Dispersions*, Academic Press, London.
18. Alford, N. McN, Birchall, J.D. and Kendall, K. (1987) High-strength ceramics through colloidal control to remove defects. *Nature*, **330**, 51–3.
19. Philipse, A.P., Bonekamp, B.C. and Veringa, H.J. (1990) Colloidal filtration and (simultaneous) sedimentation of alumina and silica dispersions: influence of aggregates. *J. Am. Ceram. Soc.*, **73**, 2720–27.
20. Novich, B.E. and Pyatt, D.H. (1990) Consolidation behavior of high-performance ceramic suspensions. *J. Am. Ceram. Soc.*, **73**, 207–12.

21. Kerkar, A.V., Henderson, R.J.M. and Feke, D.L. (1990) Steric stabilization of nonaqueous silicon slips: I, control of particle agglomeration and packing. *J. Am. Ceram. Soc.*, **73**, 2879–85.
22. Velamakanni, B.V., Chang, J.C., Lange, F.F. and Pearson, D.S. (1990) New method for efficient colloidal particle packing via modulation of repulsive lubricating hydration forces. *Langmuir*, **6**, 1323–5.
23. Young, A.C., Omatete, O.O., Janney, M.A. and Menchhofer, P.A. (1991) Gelcasting of alumina. *J. Am. Ceram. Soc.*, **74**, 612–18.
24. Lidén, E., Persson, M., Carlström, E. and Carlsson, R. (1991) Electrostatic adsorption of a colloidal sintering agent on silicon nitride particles. *J. Am. Ceram. Soc.*, **74**, 1335–9.

CHAPTER 4

Dry pressing of ceramic powders

D. Bortzmeyer

4.1 INTRODUCTION

Dry compaction is one of the most popular shape forming processes, since it involves a relatively simple technology while allowing high production rates. However, our understanding of this process is still empirical. Most industrial problems in this area are solved (if at all) by trial-and error; this is the reason for the many papers now published each year about compaction. The aim of this chapter is to review these papers.

Section 4.2 gives a short overview of compaction technology, and describes a few classical industrial problems. Sections 4.3, 4.4 and 4.5 will try to link these problems with the theoretical studies published in the literature.

Most of these results concern what might be called the 'classical' approach to compaction: pressure/density relationship, radial pressure coefficient and wall friction coefficient. The influence of powder characteristics, binder content and process parameters on these factors has been described widely in the literature and is reviewed in section 4.3.

These methods allow a description of the behaviour of a homogeneous sample, but to predict quantitatively the stress and density variations in a shaped mould, the use of continuum mechanics and computer simulation is necessary. The purpose of section 4.4 is to introduce the reader to the basic concepts of continuum mechanics, and to describe some results concerning this increasingly important research field.

Section 4.5 is dedicated to the microstructure of green parts. Granule behaviour is one of the most important factors affecting the microstructure; pore size distribution and sample anisotropy will also be discussed. A computer simulation will give some original results about the behaviour of the particles during densification.

Ceramic Processing. Edited by R.A. Terpstra, P.P.A.C. Pex and A.H. de Vries.
Published in 1995 by Chapman & Hall, London. ISBN 0 412 59830 2

4.2   INDUSTRIAL PROBLEMS

The aim of this section is to explain briefly the main features of the industrial compaction process, and to consider a few typical problems in order to link them with the theoretical aspects described in the following sections.

4.2.1   Compaction technology

(a)   Isostatic compaction

In isostatic compaction, a powder is poured into a rubber bag and stress is applied by means of a liquid which acts as a pressure transmitter. In the 'wet bag' method, the powder is poured into the bag which is submerged in the liquid (Fig. 4.1). After compaction, the bag is withdrawn from the liquid and opened to remove the part. This method is suited to large pieces, but it does not allow high production rates. In the 'dry bag' method, the rubber bag is part of the equipment. The pressure is applied by a liquid on the sides of the sample, and by a punch on the top and bottom. As a consequence, the stress is 'less hydrostatic' than in the former method. In the dry bag method, however, the filling of the mould and the ejection of the sample may be automated. High production rates are thus possible, for small pieces with relatively simple shapes. Spark plugs are a classical example of dry bag compaction [1].

(b)   Uniaxial compaction

In uniaxial compaction, the stress is applied by a punch in a mould whose side walls cannot move. Since no rubber is involved, the control of green dimensions is better than in the isostatic case. This process allows the fabrication of rather complicated shapes, even with screws or holes perpendicular to the compaction axis. In the former case, the punch rotates and translates at the same time. In the latter, the mould contains a complex system of gears and drawers. The uniaxial compaction process allows very high production rates.

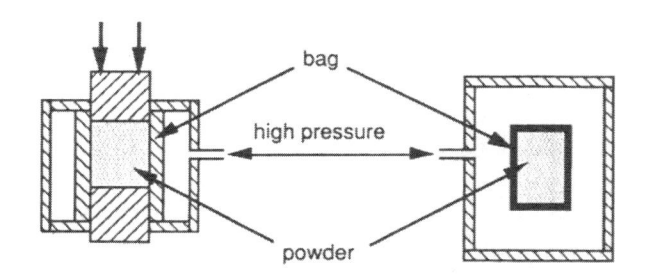

Fig. 4.1 Isostatic compaction: dry bag (left) and wet bag (right).

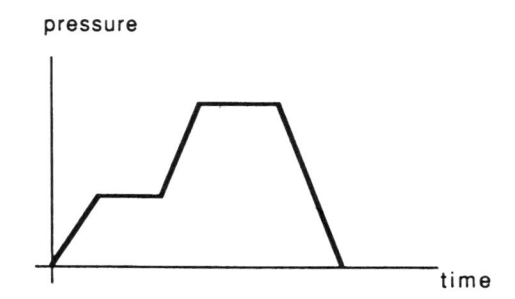

Fig. 4.2 Pressure cycle improving the green density.

Industrial pressing machines usually offer some options which are thought to improve the green sample characteristics. For example, the overall density may be increased by the application of a 'second pressure' cycle (Fig. 4.2). This feature will be discussed further.

4.2.2 A few compaction problems

(a) Density variations

Density variations in the green product will cause differential shrinkage and crack initiation upon sintering. In order to avoid this, stress and deformation variations in the sample should be minimized.

In isostatic compaction, the origin of stress variation is friction against the rubber mould or the mandrel [2]. The importance of the mould is not to be underestimated since ceramic powder deformation is very sensitive to small differential stresses (see section 4.4). As a consequence, the mould thickness and rigidity should be kept as low as possible.

In uniaxial compaction, the first important cause of stress variation is friction against the walls. It can be lowered by proper lubrication of the mould, the lubricant usually being a component of the binder. A second important cause is variations of volumetric ratio in the sample. Figure 4.3 shows a mould

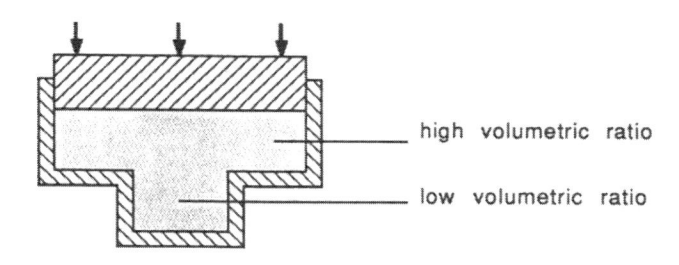

Fig. 4.3 Compaction of a mould with variable cross section.

with variable cross section. In the case of single effect compaction, the external ring of the sample will be more densified than the central cylinder. This problem is avoided through complex mould movements, to keep the volumetric ratio as constant as possible. Computer simulation can be used to predict the density variations in a mould: a few examples will be given in section 4.4.

(b) Dimensional control

To achieve near net shape forming, and thus to reduce machining costs, the green shape and dimensions must be kept reproducibly within precise tolerances.

The main problems occur with wet bag isostatic pressing. For example, consider the case described in Fig. 4.4. During the compaction of this conical sample, the powder is likely to flow along the mould (arrows), causing variations in thickness along the length of the sample [2].

Shape control is particularly difficult near corners; in order to keep a perfect 90° angle, the mould rigidity should be equal to the powder rigidity. This is obviously difficult to achieve since the powder modulus will change dramatically during densification.

Dimensional control may also be a problem during uniaxial compaction since the green sample experiences an important springback upon unloading (sometimes several percent). The exact nature of this phenomenon will be discussed in sections 4.3 and 4.4.

(c) Fracture upon unloading

Sample fracture is a very common problem, either in isostatic or uniaxial compaction. Several fracture morphologies are possible, depending upon mould shape and powder characteristics (Fig. 4.5). End capping (Fig. 4.5a) occurs mainly at high pressure. Fine powders are usually subjected to lamination, i.e. formation of sheets upon ejection due to air pressure buildup during compaction (Fig. 4.5b). After compaction of a powder without binder, the author has observed conical fracture as shown in Fig. 4.5c. Rubber relaxation after isostatic compaction of a cylinder may lead to a fracture as shown in Fig. 4.5d.

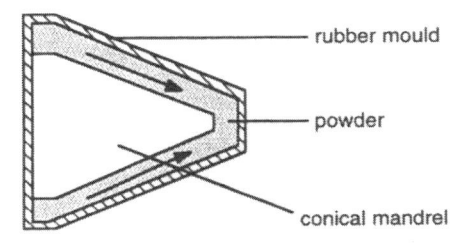

rubber mould

powder

conical mandrel

Fig. 4.4 Isostatic compaction of a conical shape with a mandrel.

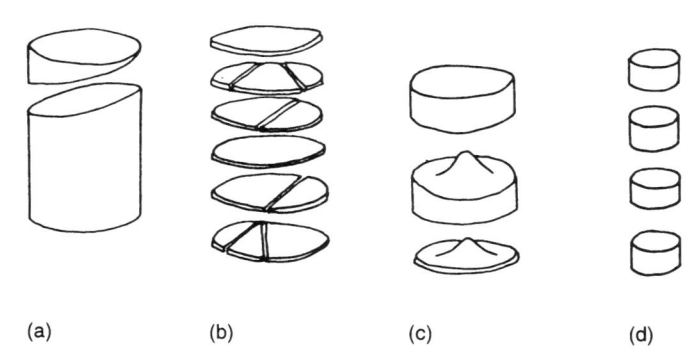

(a)   (b)   (c)   (d)

Fig. 4.5 Fracture morphologies: a = end capping; b = lamination; c = formation of cones; d = fracture of an isostatically pressed cylinder.

To overcome these problems, one may either find and eliminate the cause of tensile stresses, or increase the tensile strength of the green sample. A theoretical approach to these questions will be examined in the following sections; however, a complete analytical treatment is still lacking. Solutions are often found through trial and error among the following possibilities.

1. Internal pressure due to the air included in the sample. The typical volume change of a compacted powder being 50%, the entrapped air pressure may reach about 2 bar. Considering the low permeability of the green structure, this pressure will remain a few seconds after unloading. Although small, it is sufficient to overcome the tensile strength of green products and cause lamination. This problem may be overcome in several ways:

   (a) de-airing the powder before compaction;
   (b) optimizing the compaction rate. In isostatic or uniaxial compaction, slowing the unloading rate when the pressure is below 10 bar may give the entrapped air time enough to leave the sample without generating stresses;
   (c) in uniaxial compaction, a common solution is to eject the sample while keeping a small pressure on it until the air has escaped ('ejection pressure'). Thus, the overall stress is always compressive. Of course, the applied pressure must be lower than the free compression strength of the sample.

2. Tensile stresses upon unloading arising from stress variations during loading. The stress variations caused by the compaction pressure may become tensile upon unloading (this has been shown by numerical computation). This is particularly true in uniaxial compaction, near the upper corners. The solution is once again to keep a small load over the sample while it is ejected, and to minimize stress variations by a proper mould lubrication or mould design.

3. Friction stresses on the mould during ejection. The slip-stick phenomenon sometimes observed during the ejection of a uniaxially compacted sample is obviously detrimental to the sample quality. The vibrations are likely to cause crack initiation and propagation. This can also appear in isostatic compaction, if the rubber bag slips against the powder [2]. Careful attention to mould smoothness and lubrication may help to solve this problem.

4. Powder humidity. If the powder is too humid, it may stick to the punches (uniaxial compaction) or the mandrel (isostatic compaction). Drying the powder will usually solve the problem. But if the powder is not humid enough, the green compact is likely to reabsorb atmospheric water after ejection. Sometimes the dimensional change associated with this phenomenon is large enough to break the sample. This is particularly true with large samples and hydrophilic material, for example clays.

These few examples certainly do not fully describe the very important and complicated question of sample fracture. Experience, and trial-and-error methods, are still the best ways to solve this problem.

4.3 MACROSCOPIC BEHAVIOUR: THE CLASSICAL APPROACH

Dry compaction is used in powder metallurgy, ceramic powder processing and for pharmaceutical excipients. The characteristics of these powders are rather different, but the same concepts can be applied. That is why it is often interesting to examine the results from these other fields.

The compaction behaviour of a powder, as described in the literature, may be summarized by four factors: pressure/density relationship; radial pressure coefficient; wall friction coefficient; green tensile strength. This section is intended to describe the main features of this 'classical' approach, including theoretical models, experimental methods and results.

4.3.1   Density/pressure relationship: theoretical and experimental

Ceramic powders are often compared on the basis of their pressure/density relationship. As a first step, we will examine the evolution of density during compaction. The evolution of the packing microstructure will be reviewed in section 4.5, which is dedicated to the microscopic study of compaction.

Before examining the published results, it is worth beginning with some results concerning the micromechanics of compaction. A deeper insight into this question will be the subject of section 4.5, but this will help us to understand the macroscopic behaviour of the powder.

(a)   Micromechanics of powder compaction

The first point of interest is the factors that make ceramic powder density very low. The relative density of a packing of cohesionless monosized particles is about 0.63; it is increased for polysized particles. However, the cohesion force of ceramic particles (van der Waals force) is far greater than their weight: for submicron zirconia particles, for example, they are respectively about 10^{-7} N and 10^{-13} N. As a consequence, particles adhere to each other instead of falling one at a time, and this explains the porosity of the initial powder packing. A simple calculation, however, allows the determination of the mean interparticle force resulting from macroscopic stress: for a few MPa applied on the sample, it becomes far more important than the van der Waals force. Thus, the packing porosity under high pressure is not caused by interparticle cohesion, but by geometrical structure: the particles are arranged in vaults that resist the compaction pressure. Densification is the result of the progressive destruction of these vaults.

This mechanism was shown simultaneously by L. Kuhn [3], with experiments on barrels, and by the author [4] with a numerical simulation calculating the behaviour of a collection of discs. The experiment of Kuhn is very interesting: several barrels (without cohesion) are poured in a box, and a few pores are created by removing some of them. Axial pressure is applied on the mould with an upper punch: it is shown that the pore closure proceeds through abrupt buckling of the barrel structure. Section 4.5 will show that a numerical simulation is able to reproduce this phenomenon.

(b)   Experimental

It is worth examining the experimental details of a pressure/density relationship determination, since on one hand some precautions are to be taken to avoid an erroneous interpretation of the results, and on the other hand a compaction experiment may give far more information than a simple density measurement.

The pressure/density relationship determination in isostatic compaction is difficult to perform, since at high pressure the liquid is usually more compressible than the powder. Moreover, a complete de-airing of a high pressure cell is not easy, and the presence of air bubbles is of course detrimental to the density measurement. Uniaxial compaction, on the other hand, is rather simple: the powder is simply poured in the mould, the density being measured geometrically inside the mould, or outside after ejection (of course, a measure inside the mould must be corrected by a blank experiment [5]). Most of the published results concern uniaxial compaction.

In uniaxial compaction, the stress homogeneity is not as good as in the isostatic case. That is why the aspect ratio of the compact should be made as small as possible, preferably less than 0.3; otherwise, one will measure the wall friction coefficient as well as the true pressure/density curve. For the same

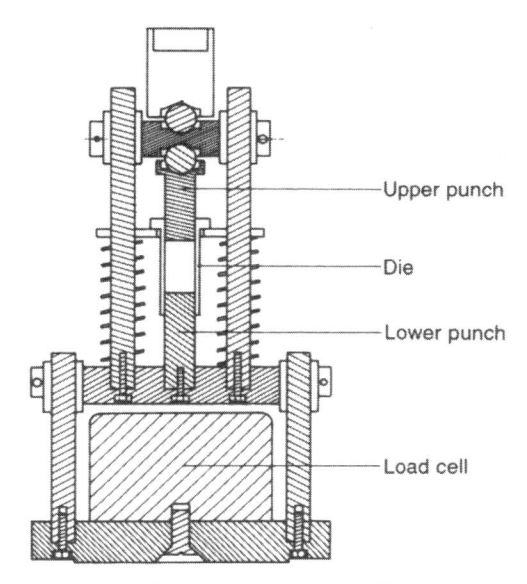

Fig. 4.6 Compaction cell.

reason, the cell should be preferably 'double effect'. An example of such a device is given in Fig. 4.6.

The comparison between the densities measured inside or outside the mould gives information about the green relaxation (Fig. 4.7). This is important because the green strength may be strongly dependent on residual stresses.

Attention should be paid to the time interval between compaction and density measurement. With a few binder formulations, a slow stress relaxation may occur and the density decrease by several percent over a few hours after ejection.

4.3.2 ρ/P relationship: influence of powder characteristics

(a) Introduction

Since the early work of Lukasiewicz [6], ceramists have commonly used the semi-logarithmic plots of the pressure/density relationship (Fig. 4.7). Sometimes, the slope of the pressure/density curve in a semi-logarithmic plot changes at the so called 'breakpoint pressure'. Several authors have shown that this breakpoint is a measure of the granule or agglomerate hardness [7–10]: a scanning electron microscopy (SEM) examination of fracture surfaces before the breakpoint shows that the granules are not destroyed and that the densification mechanism is granule rearrangement. After the breakpoint pressure, however, the granules show extensive plastic deformation.

Dry pressing of ceramic powders

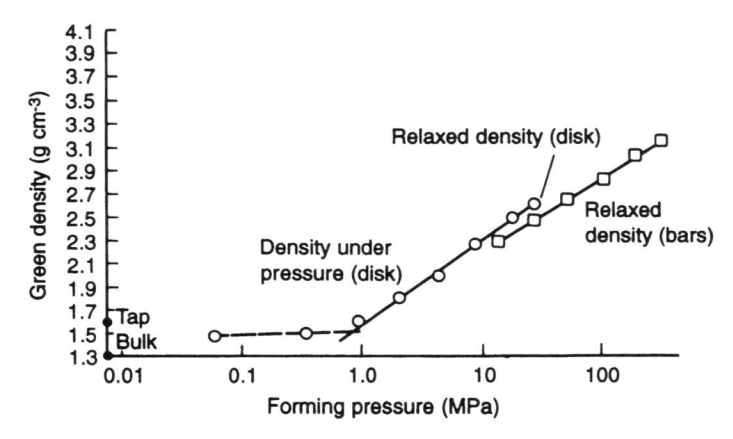

Fig. 4.7 Effect of forming pressure on green density for spray-dried Mn–Zn ferrite with 1.5% PVA binder. (From: Binder systems in ferrite, J.W. Harvey, D.W. Johnson, *Am. Ceram. Soc. Bull.*, **59**, 637–645 (1980), reprinted by permission of the American Ceramic Society.)

It must be recalled, however, that the precise determination of the break-point is difficult since continuous *in situ* measurements show a progressive change in the slope, rather than a breakpoint [4, 11]. Moreover (and this will be important for the green microstructure), the shape of the granules may be still distinguishable for pressures far greater than the breakpoint [12].

There has been much effort to demonstrate this semi-logarithmic relation from theoretical considerations. For example, Caligaris *et al.* [13] tried to use a 'thermodynamic' demonstration involving the use of an entropy measure of the packing disorder. But such an approach is difficult, since an exact description of the 'disorder' of the packing is not self-evident. The coordination number of the particles, for example, is certainly an insufficient measure. The reader interested in these theories might read the work by Kanatani, [14]. This author shows that the application of statistical thermodynamics, even for the relatively simple case of two dimensional packing, is quite difficult.

Moreover, one must keep in mind that a lot of other relations have been proposed. For example, Kawakita's equation [15]

$$\frac{V - V_0}{V_0} = \frac{aP}{1 + bP}$$

is often used in ceramics, since it predicts a finite density at infinite pressure (V_0 is the initial volume of the sample, V the volume at pressure P, a and b are empirical parameters). Heckel's relation [16] is well known in pharmaceutical technology

$$\log\left(\frac{1}{\varepsilon}\right) = bP + a \quad \text{where } \varepsilon \text{ is the porosity.}$$

Chaklader [17] gives a list of several other laws used in the literature.

(b) Ungranulated powders

The main characteristics of ungranulated powders are mean particle size, size distribution, morphology and aggregation state. The pharmaceutical powders are also often compared in terms of particle hardness.

When the **mean size** of an ungranulated ceramic powder is modified, one observes a shift of the pressure/density relationship, Fig. 4.8.

This result has also been obtained by Carless [18] with an aspirin powder. It is related to the microscopic mechanism of compaction described above. The initial density of the powder depends on the ratio particle weight/interparticle cohesion and increases with the particle size. But the interparticle cohesion becomes rapidly negligible compared with the forces induced by the overall pressure, so that the mechanism of densification (buckling of vaults) becomes independent of particle size. Thus, above a few MPa the compaction rate is the same for all powders.

This mechanism sheds light on the importance of the initial density of the powder. This is particularly important for the choice of the granulation conditions.

The influence of the **size distribution** of powders has been investigated by Zheng and Reed [19]. A powder with a wide size distribution is expected to have a higher maximum density than a monosized powder (section 4.5). As a consequence, when the two powders have the same initial density, the slope of the pressure/density relationship is greater for the polydisperse population.

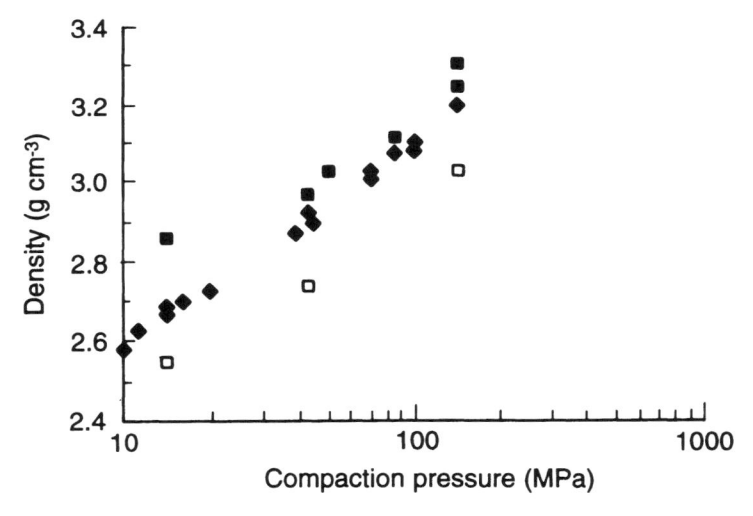

Fig. 4.8 Pressure/density relationship for three powders with different mean sizes (□: 0.33 μm; ◆ : 0.53 μm; ■: 0.92 μm); (source: Bortzmeyer 1990, ref. [4]).

This shows that our explanation concerning the influence of the mean particle size is only an approximation.

The influence of **particle morphology** on the flow properties of powders (taped and bulk densities, angle of repose) has been widely studied for cohesionless powders. But it is difficult to find ceramic powders differing only in morphology; that is why an extensive study of this parameter is still lacking, even if some very interesting articles have been published. For example, Leiser and Whittemore [20] studied the progressive fracture of alumina particles with high aspect ratio. Yamaguchi and Kosha [21] showed that the porosity of a compacted platelet-like powder is very different from the porosity of a spherical powder.

The problem of **aggregated powders** is of primary industrial importance. Aggregates are often created during the calcination step of powder synthesis, and their hardness depends on calcination temperature, synthesis solvent, particle size and reactivity. Aggregation is very detrimental to the sintering process, so it is important to investigate its influence on density and microstructure.

As was stated before, the pressure/density relationship often displays a so-called 'breakpoint' characterizing the beginning of aggregate (or granule) fracture [7, 94]. The harder the aggregates, the higher the breakpoint pressure (Fig. 4.9). However, the breakpoint is not always observed; large, monosized, homogeneous aggregates are probably necessary for this phenomenon to occur. Ciftcioglu *et al.* [22] conducted an extensive study of the effect of aggregate hardness, both on green and sintered samples. They found the pressure/density relationships to be parallel, shifted to higher compaction pressures when the aggregates are harder.

When the compaction involves the plastic deformation of particles (this is the case with pharmaceutical excipients), **particle hardness** becomes important. Several studies show the influence of particle hardness [18], or the morphological evolution of particles during compaction [23, 24]. Pharmaceutical excipients may also undergo some sintering between particles during compaction, while this phenomenon is not likely to occur with ceramic powders.

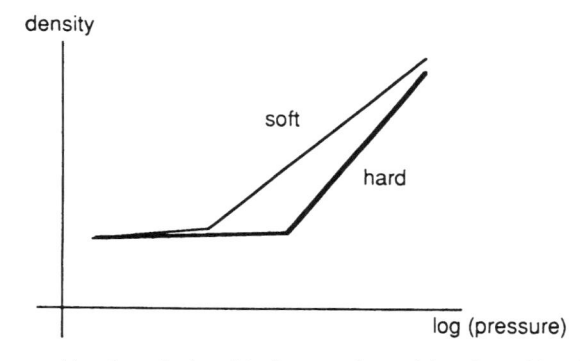

Fig. 4.9 Pressure/density relationship for powders with soft and hard aggregates.

(c) Granulated powders

So far, we have examined the influence of particle characteristics in an ungranulated powder. However, most powders are granulated before compaction in order both to improve their flow and to mix them with a convenient binder. The effect of the binder is of primary importance on the compaction results, but the influence of the granule structure is also (and may be more) important. Therefore the compaction behaviour of a granulated powder without binder will be presented first.

The reader interested in the spray drying process might read, for example, references [25–27].

The initial density of a granulated powder depends on its granule density (ρ_g) and granule size distribution. For monosize free flowing granules, the initial density is expected to be close to 0.6 ρ_g since 0.63 is the relative density of the 'random close' packing of monosized spheres. This initial density will influence the whole pressure/density relationship. Consider, for example, a given powder granulated without binder to give two populations A and B, with $\rho_g(A) < \rho_g(B)$. Zheng and Reed [19] have shown that the density of A (46.3% granule density in Fig. 4.10) is always smaller than the density of B (61.5% granule density in Fig. 4.10).

On the basis of this result, one might think that having a high granule density is better for the green quality. However, if the granules are too dense they may be too hard and the compaction pressure will not destroy them. There may remain some large pores which will generate flaws upon sintering. For example, the green compacts from powder B are indeed denser than those from A, but they are also less dense than the initial granules (the granule densities of powders A and B are shown on Fig. 4.10).

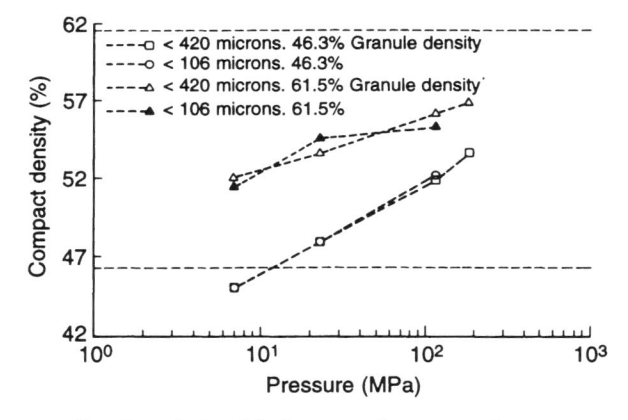

Fig. 4.10 Pressure/density relationship for a powder prepared at two granule densities. (From: Particle and granule parameters affecting compaction efficiency in dry pressing, J. Zheng, J.S. Reed, *J. Am. Ceram. Soc.*, **71**, C.456–C.458 (1988), reprinted by permission of the American Ceramic Society.)

We will now turn to the question of the choice of binder. The binder composition is usually the 'private recipe' of each ceramist; it is kept secret, and only some general information is available in the literature. An extensive classification is given by Bast [28]. Two main types of binder have been fully described in the literature: those based on classical hydrosoluble polymer (mainly poly(vinyl alcohol) PVA and poly(ethylene glycol) PEG) and those based on synthetic emulsion polymers (usually called latexes).

The choice of a binder depends on a lot of parameters, including pyrolysis behaviour, rheological behaviour and economic considerations. Our discussion will be limited to the compaction behaviour. The binder influences granule strength, granule density and particle sliding properties. These three parameters will in turn modify the pressure/density relationship.

The question of **granule strength** and structure is very important; if the granules are too hard, they are not destroyed by the compaction pressure and the green compact microstructure is not homogeneous.

As a consequence, a great deal of work has been done to measure directly the crushing strength of the granules. Granule testing machines are described for example in references [29–31]. The first author shows that the choice of the granule behaviour (brittle or plastic) depends on the compaction pressure. The two latter authors were able to detect the presence of a hard shell around the granules over a certain size. These methods are indeed very powerful; but they are also very time consuming, and to our knowledge no extensive study of the correlation between granule crushing strength and compaction behaviour has been reported. Usually, the granule strength is estimated through the breakpoint pressure, or by an examination of compact fracture surfaces.

Several articles have been devoted to the relation between binder hardness and granule cohesion. The simplest measure of a binder hardness is the Shore hardness of a cast binder film. It seems that there is a direct correlation between this characteristic and granule cohesion [12]. A more indirect determination is the binder 'glass transition temperature' T_g, i.e. the temperature separating its 'crystalline' and 'amorphous' phases. Above T_g, the binder is soft and ductile; below, it is hard and brittle. The results show that the granules are indeed softer when the pressure temperature is above T_g. For a given pressing temperature, the higher T_g, the harder the binder and the more cohesive the granules [8].

This is the reason for the use of 'plasticizers' which lower the T_g of a given binder (PEG, for example, is a plasticizer for PVA). A precise compromise must be found between binder hardness (which is important for green strength) and plasticity (which allows good granule destruction). This study is complicated because humidity is a good plasticizer for PVA; as a consequence, the compaction results of a PVA-based formulation will depend on the humidity of the atmosphere [9, 11].

It is important to note that binder is not uniformly distributed inside granules. The mechanism of granule drying is well known [25, 26]: the water flows from inside the granule to be vaporized at the surface. Thus, the binder migrates with the water and tends to accumulate near the surface, forming a hard shell which is detrimental to the green microstructure; as stated before, a direct measure of granule crushing strength may reveal this problem. The main value of latexes, compared to PVA for example, is that the relatively large size of their particles prevents them from migrating with the water.

Note that an interesting method for granule structure examination has been described by Uetsamu [33].

Granule density is one of the most important parameters for granule strength, and particular attention should be paid to it. It will depend on the spray drying conditions (solid content of the slurry, drying temperature, and also dryer design), but also on the binder system itself, through complicated mechanisms (viscosity, diffusion rates, etc). An extensive review of the equations describing droplets drying in a spray-drier may be found in reference [25]. However, the numerical parameters of these equations are often unknown, so experiments are preferable for the determination of optimal process conditions.

The slope of the pressure/density relationship will depend both on the **binder lubrication properties**, and on the initial powder density. As a consequence, no general rule is available: the slope will depend, for example, on the spray-drier used for granulation. The best way to examine the lubrication properties is to measure the radial pressure coefficient of the powder.

4.3.3 ρ/P relationship: influence of process parameters

(a) Loading

For a given powder and binder system, the pressure/density relationship will be affected by two pressing parameters: punch speed and loading sequence.

The most important parameter in industrial applications is the **punch speed**. An increase in speed will usually be detrimental to density because the binder has a viscous behaviour, thus the ability of particles to re-arrange is reduced [34, 35]. Moreover, it will have a detrimental effect on sample quality: entrapped air will not escape fast enough, which will result in pore pressure buildup and possibly fracture upon unloading and ejection.

Time is important not only for density improvement but also for microstructure improvement. Naito [36] described the influence of stress relaxation on pore size distribution.

Another parameter of interest is the **loading sequence**. It is usually a monotonic loading followed by a monotonic unloading. However, it has been shown that the green density rises when it is submitted to a cyclic loading: it is a logarithmic function of the number of cycles [37]. This behaviour is a direct consequence of the microscopic mechanism of compaction, which was described earlier, that is, the buckling of the microscopic structure. The packing resists densification because it is organized in vaults; this structure is more efficiently destroyed by a cyclic loading than by a simple pressure increase. It has been demonstrated that the cyclic application of a pressure of 40 MPa could be as effective as a static pressure of 600 MPa [38].

To our knowledge, this interesting phenomenon is not used in industrial equipment. However, as we already mentioned, some compaction equipments are able to apply a 'second pressure' on the sample. Although it has been shown [38] that a load cycle is more efficient when the bias pressure is zero, this second pressure may improve green density and microstructure.

Another example of the positive influence of cyclic load is the use of sonic or ultrasonic waves to help compaction. Emeruwa *et al.* [39] have shown that the application of ultrasonic waves at a given pressure increases the green density and improves the green structure (granule destruction).

(b)   Unloading and ejection

A lot of problems encountered in shape forming by compaction occur during unloading. However, this part of the process is often neglected in compaction studies, or briefly referred to as the 'elastic springback'. We would like to stress the importance, both theoretical and technological, of this part of the process.

A first important point is the real nature of the green deformation during unloading. It is not purely elastic but always contains a plastic component, which may be quite important. Otherwise, the increase of density upon cycling could not exist. The order of magnitude of springback is indeed substantially larger than a classical elastic deformation: inside the mould, it may reach several percent in the axial direction (which is relaxed first). As a consequence, there is a great difference between the pressure/density curves inside and outside the mould.

The microscopic causes of this phenomenon have long been studied in soil mechanics. We will see later (section 4.5) that the movements are not homogeneous among the particles; even a macroscopic isotropic deformation occurs through complex rearrangements at the particle level. Nor are the interparticle forces homogeneous: the macroscopic stress is transmitted from one plunger to another through a complex network of loaded particles, directly affecting only a small number of them. Upon unloading, stress relaxation similarly cannot occur through a simple elastic homothetic deformation. It necessarily involves some plastic sliding, which occurs readily since the energy of adhesion between particles is quite weak compared with the elastic energy stored under loading.

The particle morphology has a great influence on this process; for example, particles with a high aspect ratio, or highly agglomerated, are likely to store a high amount of bending energy upon loading, which will cause a greater amount of springback.

The second important phenomenon arising during unloading concerns stress variations in the sample during ejection. This question will be examined at length in section 4.4, and it will be shown that stress variations arising during loading may turn into tensile stresses upon unloading.

The loading rate may be also important during unloading. This is particularly true in isostatic compaction, as was mentioned earlier (section 4.2). In that case, two phenomena are dangerous for the green structure. First, the pore pressure due to entrapped air: as in the case of uniaxial compaction, the pressure must be released at a sufficiently low rate, particularly at less than 10 bar. Second, it is thought that the relaxation of the rubber mould, and its sliding against the green body, may also cause tensile stresses and fracture. The unloading rate, as well as the mould design, may be an important parameter for the control of these problems.

Last but not least, the viscoplastic behaviour of the binder may cause a slow relaxation of stress after ejection. This is important for the accurate measurement of the pressure/density relationship, because the sample dimensions (and density) may change by several percent. It is also important for technological applications: for example, the green compacts should not be machined before complete relaxation has occurred. Otherwise, sample fracture is more likely to occur.

4.3.4   Radial pressure coefficient

When a powder is uniaxially compacted by a punch (pressure σ_z), part of this pressure (σ_r) is transmitted on the matrix walls. The radial pressure coefficient $k = \sigma_r/\sigma_z$ is an interesting characteristic since it describes the powder's 'fluidity'. If k is a constant equal to 1, for example, the powder behaves like a fluid; uniaxial compaction is equivalent to isostatic compaction.

This fluidity coefficient is an indication of the ability of the powder to transmit stress evenly in a shaped sample (Fig. 4.11). We will see, however, that the wall friction coefficient is also important in the determination of stress and density distribution.

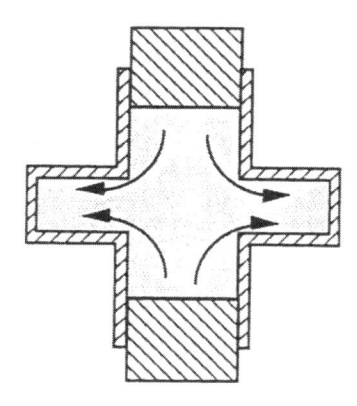

Fig. 4.11 Compaction of a shaped sample: the corners are better filled if the powder is more fluid.

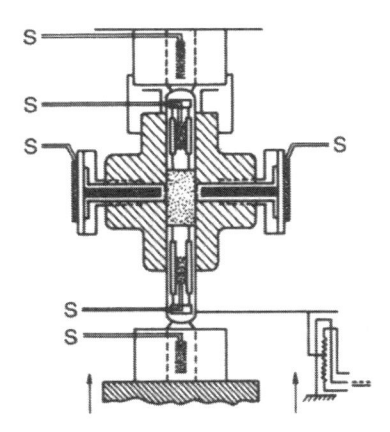

Fig. 4.12 Apparatus used to measure the distribution of the normal stress on die wall and punches during compaction: S = pressure gauges. (From: S. Strijbos, P.J. Ramkin, R.J. Klein Wassink, J. Bannick, G.J. Oudemans, Stresses occuring during one-sided die compaction of powders. *Powder Technol.*, **18**, 187–200 (1977), reprinted by permission of Elsevier Sequoia S.A.)

(a) Experimental

Two different methods for measuring the radial pressure coefficient have been described in the literature. The most direct method consists in applying directly a pressure gauge on the powder, through the matrix (Fig. 4.12). It has been used for example by Strijbos [40].

Note that the radial pressure is not homogeneous along the sample height. A valid comparison between powders must be done with a constant sample height (as small as possible), and at a constant position of the radial gauge.

The second method consists in measuring the matrix deformation due to the radial pressure (Fig. 4.13).

In this method, the matrix is not disturbed by the measurement because its deformation is very small (< 0.1%). But even if the radial pressure is constant along the sample height, the result depends on the gauge's position on the matrix (this is obvious in Fig. 4.13). Di Milia and Reed [32] have made an extensive study of this problem. The gauges must therefore be always at the same position, and the measurement is valid only for a given sample height; the double effect configuration is not possible.

(b) Results

Within a good approximation, the radial pressure is a linear function of the axial pressure (Fig. 4.14) [18, 40, 41]. In other words, the radial pressure coefficient is a constant for a given powder.

Table 4.1 compares the radial pressure coefficients of different zirconia powders, with varying sizes, morphology and binder content. The mean

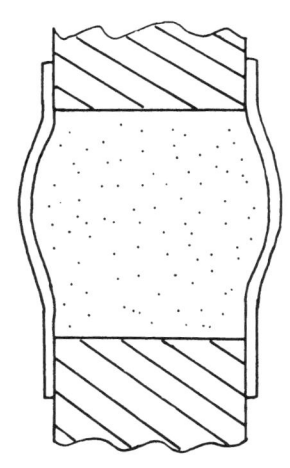

Fig. 4.13 Matrix deformation due to the radial pressure (exaggerated).

Table 4.1 Radial pressure coefficient for different zirconia powders (range 0–30 MPa). At high pressure, the polymer film between the particles is broken, lowering the radial pressure coefficient to its binderless value

	Radial pressure coefficient, k
Powder size (μm)	
0.33	0.406
0.53	0.429
0.92	0.392
Powder morphology	
irregular	0.406
spherical	0.500
Binder content	
none	0.429
4% PVA	0.488
4% PVA, 2% stearic acid	0.498

particle size has a small influence, while particle morphology and binder have a greater effect.

It is sometimes stated that the radial pressure coefficient is more strongly affected by wall lubrication than by binder content. This is not exactly true: the radial pressure coefficient is a powder property, which cannot be affected by a modification of the walls. But a bad wall lubrication may cause large stress variations along the walls; thus, the measured radial stress may be lower than the stress that would exist in a sample with even stress distribution.

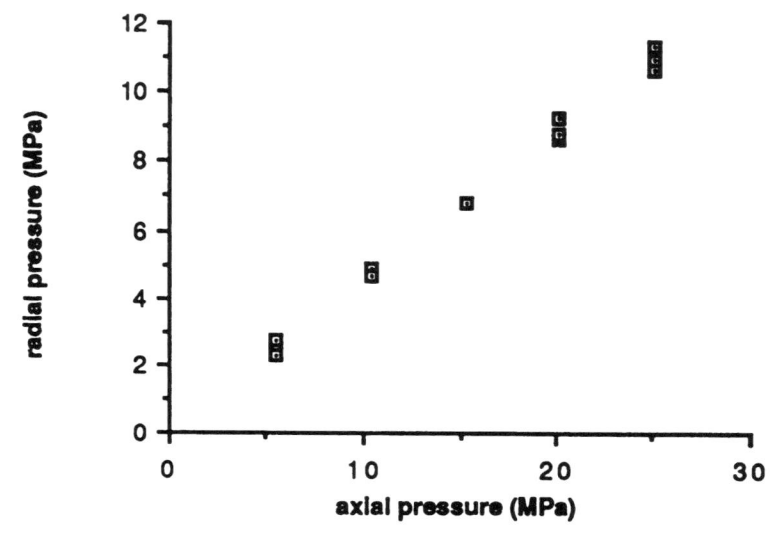

Fig. 4.14 Radial pressure against axial pressure during the densification of a zirconice powder. (Source: Bortzmeyer 1990, ref. [4].)

4.3.5 Wall friction coefficient

As a first approximation, the powder/wall interaction may be considered to be Coulomb friction, with friction stress and radial pressure related by $\tau_r = \mu_p \sigma_r$. The wall friction coefficient, μ_p, is very important for stress and density distributions (section 4.4). However, it has scarcely been measured.

The simplest method is to measure simultaneously, during uniaxial compaction, the stress, σ_a, applied on the top punch, the stress, σ_t, transmitted to the bottom punch and the stress, σ_r, transmitted to the walls. One has

$$(\sigma_a - \sigma_t)\pi R^2 = \mu_p \sigma_r 2\pi Rh$$

where R is the mould radius and h the green height. This method has been applied by Henke [41], while Strijbos [42] has used a device specially designed for the measurement.

The main conclusion of these authors is that the wall friction indeed obeys a Coulomb law. The friction coefficient depends mainly on the wall roughness and lubrication: the binder content and composition of the powder seem to be less important. It is expected on theoretical grounds that the wall friction coefficient is higher for small particles [43]. This has been observed by the author on zirconia powders [4].

4.3.6 Green tensile strength

Sample fracture is a very important problem in every shape forming process: compaction (upon unloading), slip casting (during drying), injection moulding (during pyrolysis). In order to control it, one has to know the 'driving force' of the fracture (i.e. what are the stresses causing the rupture) and the mechanisms of cohesion in a green body.

The former is very dependent on the shape forming process. As was mentioned before, the stress inhomogeneities arising during uniaxial compaction because of wall friction turn into tensile stresses upon unloading. The friction stresses during ejection (often associated with a slip-stick phenomenon) may also contribute to the opening of cracks.

A fundamental understanding of the mechanisms of cohesion is very important, but still lacking. More and more research is being dedicated to this complicated subject, which cannot be fully addressed in this section. We would like, however, to give a brief review of the main ideas necessary to understand it.

(a) Experimental measurement of green strength

The easiest way to determine the tensile strength of green bodies is the so-called 'diametral compression test' [44, 45], (Fig. 4.15). Its results are said to be very close to those of a conventional tensile test.

However, since the green fracture is brittle, the mean strength obtained from a few measurements is not sufficient to characterize the green cohesion. A study of the Weibull modulus and the critical stress intensity factor, as a function of the shape forming process, would also be useful. This can be achieved through three-point bending experiments.

(b) Theory for green strength

Considering the rather simple structure of a green body, it is worth trying to determine the macroscopic tensile strength from microscopic factors (tensile

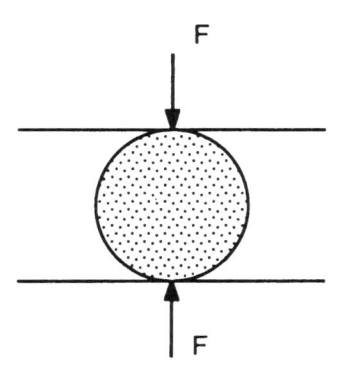

Fig. 4.15 Diametral compression test: the stress is uniformly tensile along the diameter joining the loading lines.

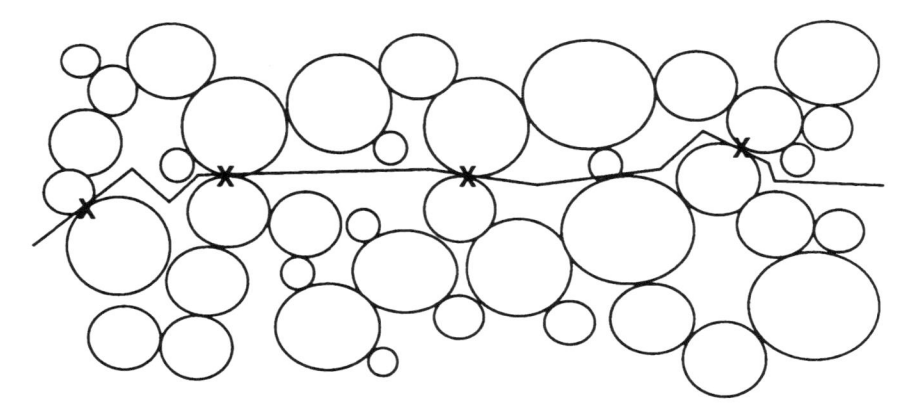

Fig. 4.16 Bonds crossing the fracture plane in two dimensions.

strength of one particle/particle bond, distribution of bonds). The simplest of these methods is to count the number of particle/particle bonds in the failure plane (Fig. 4.16).

This calculation is rather simple [46]. Then one assumes that all bonds have the same cohesion f_0 and are broken at the same time. The result for a random packing of monosized particle is:

$$\sigma_t = \frac{f_0 \rho}{4(1 - \rho)r^2}$$

where ρ is the relative density and r the particle radius.

However, this calculation is not very satisfactory since the particle/particle bonds crossing the fracture plane are not broken at the same time. It is recognized that the above equation does not always fit well the experimental results, when f_0 and σ_t are measured separately [47].

This is not surprising since it is obvious that the tensile failure of a green body is caused by crack propagation (like any brittle material). This means that the tensile strength of green bodies must be studied using the tools developed for fracture mechanics. This point has been discussed by Kendall [48] and Adams [49]; in fact, only the rupture energies (and not the cohesion forces) of the broken contacts in the failure plane can be summed. Kendall calculated analytically the tensile strength of several regular packings of monosized particles, and extrapolated the behaviour of a random one from these data. His formula is

$$\sigma_t = \frac{15.6 \rho^4 \Gamma_R^{5/6} \Gamma_E^{1/6}}{\sqrt{2rc}}$$

where c is the crack length, and Γ_R and Γ_E are surface energies measured by rupture and modulus experiments respectively.

(c)   Results

Kendall [48] found a good correlation between his own model and experimental results from three-point bending tests. The toughness measurement on submicron titanium dioxide green compacts indicated that the typical crack size was of the order of 600 μm. Adams [49] also used fracture mechanics to analyse the strength of sand mixed with binder. While the particle size was far greater than in the former case, the crack size was about the same.

Our own results with several binderless zirconia powders, using the diametral fracture test, showed that the green strength depends only on the compaction pressure whatever the density obtained [50]. Thus, the major contribution to the overall strength of binderless green bodies obtained by uniaxial compaction seems to be the mechanical interlocking of the particles. Thompson [53] used this assumption in his model of the end-capping problem. It is our opinion that the balance between the different contributions to the strength depends critically on particle shape and roughness, as well as on the forming process. As a result, sample fracture upon unloading is dependent on all these parameters.

When a binder is added to a powder, the green tensile strength will depend mainly on the binder characteristics (usually on the Shore hardness) and on the adhesion between the binder and the particle's surface. The main problem is that green strength and granule strength are often related, so that an increase in green strength is detrimental to green microstructure. Thus, the preparation of a binder requires a number of trade-offs.

4.3.7   Conclusion

We have examined four important factors that characterize the mechanical behaviour of a powder: the pressure/density relationship, the radial pressure coefficient, the wall friction coefficient, and the green tensile strength.

The pressure/density relationship characterizes the compaction response of the powder. The radial pressure coefficient and wall friction coefficient describe the interaction of the powder with the mould: they can be used for an estimation of stress variations in the mould. The green tensile strength characterizes the ability of the powder to be handled and machined. The influence of powder parameters on these factors, as published in the literature, was reviewed.

It is our opinion that measurement of these four factors is necessary, and rather sufficient, to characterize a powder and compare efficiently different powders.

4.4 MACROSCOPIC BEHAVIOUR: THE MECHANICAL APPROACH

4.4.1 Density distribution measurement

In the preceding section, we have determined the behaviour of a sample submitted to homogeneous stress and strain. However, the ceramist is concerned not only in the density that may be obtain with a given powder, but also in the density variations of this powder in a shaped mould.

The density distribution in a mould may be measured by several methods: microhardness distribution [54]; X-ray radiography [55]; NMR [56]. For example, in uniaxial single effect compaction one has the well known pattern shown in Fig. 4.17.

The density variations in a cylindrical mould may be calculated with a few assumptions [51]:

- the axial stress distribution in a plane perpendicular to the mould axis is parabolic

$$\sigma_z(r, z) = r^2 f(z) + C(z), \text{ where } f \text{ and } C \text{ are to be determined;}$$

- along the die wall (radius R), the radial stress is always proportional to the axial stress

$$\sigma_r(R, z) = k\sigma_z(R, z);$$

- the friction on the die wall obeys Coulomb law

$$\tau(R, z) = \mu\sigma_r(R, z).$$

A relatively simple calculation leads to the axial stress distribution in the mould

$$\sigma_z = Br^2 \exp\left(\frac{4k\mu z}{R}\right) - \frac{C}{R^2}$$

where C is shown to be constant.

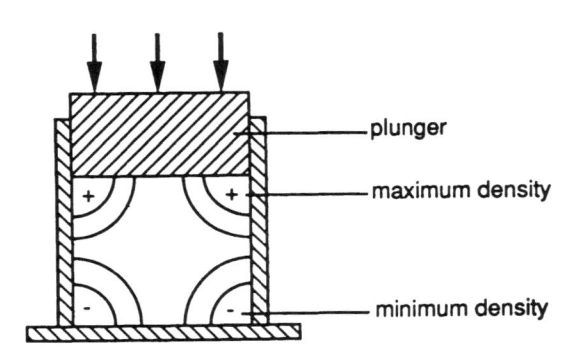

Fig. 4.17 Density distribution in a cylindrical mould under uniaxial compaction.

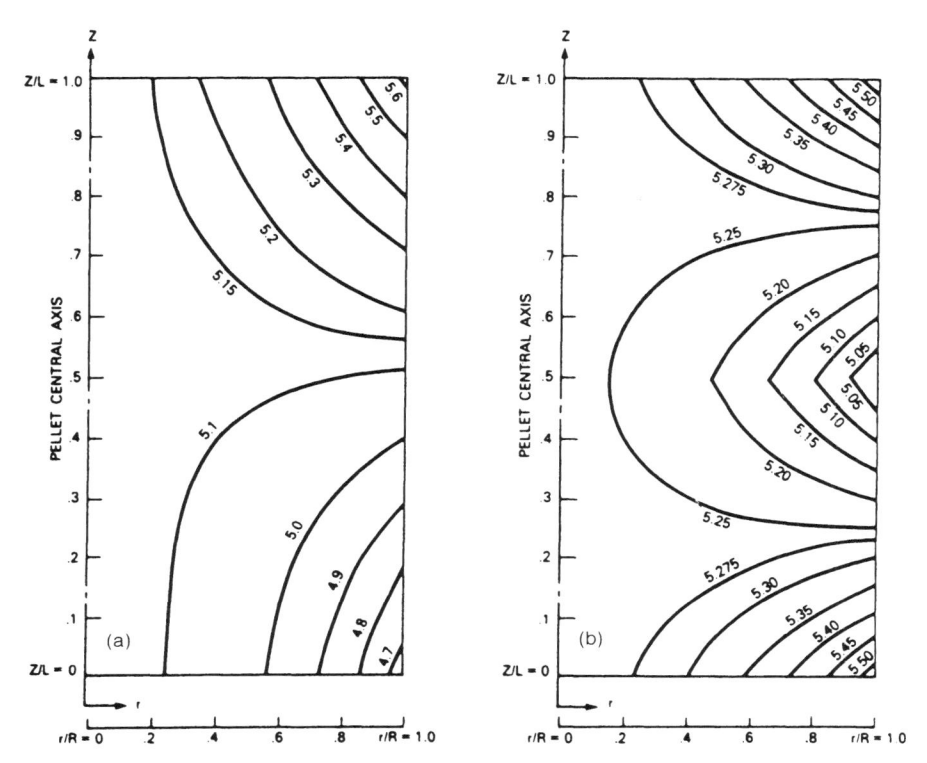

Fig. 4.18 Lines of constant density for a pellet formed in: a = single acting press and, b = symmetrical double acting press. (From: Mechanics of Powder pressing: I Model for powder densification, R.A. Thompson, *Am. Ceram. Soc. Bull.*, **60** (2), 237–243 (1981), reprinted by permission of the American Ceramic Society.)

The density distribution is readily obtained if one knows the relationship between stress and density. The author uses

$$\rho = \rho_i + T\sigma_z^{1/3}$$

Figure 4.18 is an example of this calculated density distribution; it indeed looks like the classical experimental results.

Another method has been described by Mac Leod with a radioactive powder [57]: the cylindrical compact is cut along its axis and slightly pressed against a photographic film. The amount of radiation is proportional to the density of radioactive powder, so that a direct imaging of the green structure is obtained. The patterns are rather interesting (Fig. 4.19).

The theoretical analysis sheds light on the influence of the parameter $k\mu L/R$ where L = total height of the mould. The higher this ratio, the greater the density variations along the mould height. As was stated before, an increase in k (powder fluidity) improves the green homogeneity, only if there is a simultaneous decrease in μ.

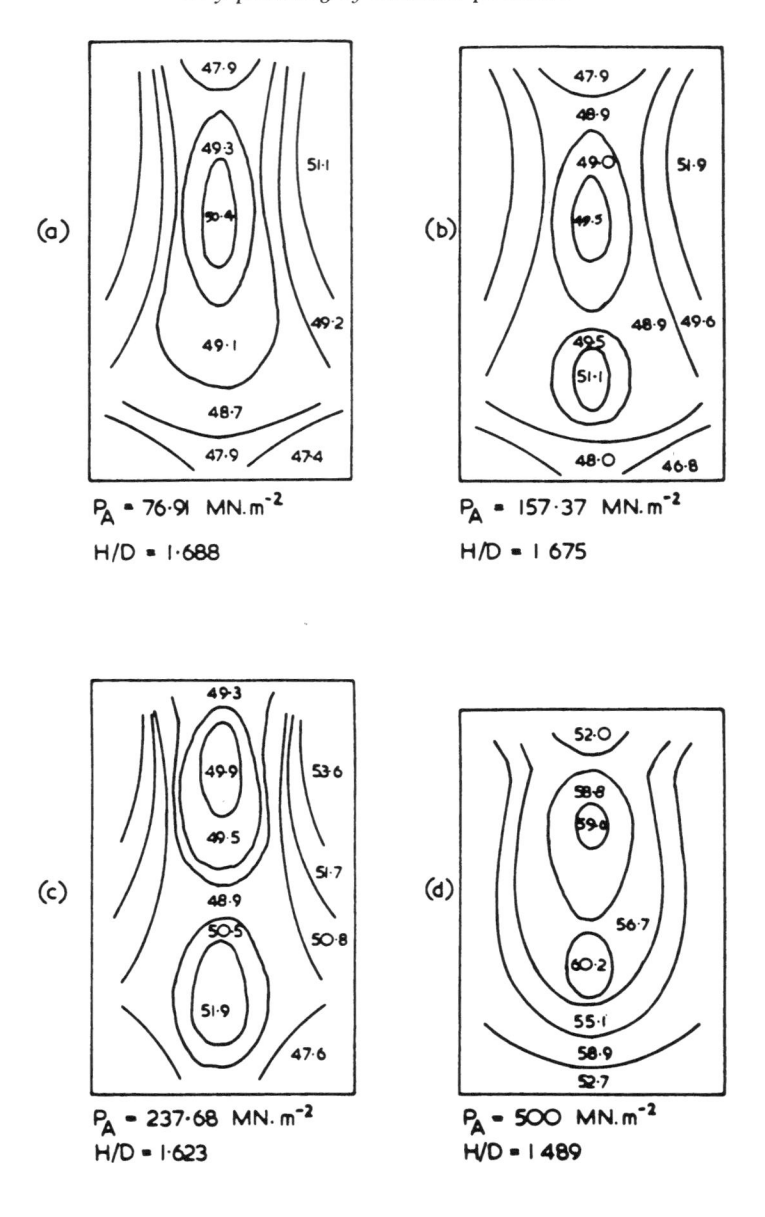

Fig. 4.19 Density patterns for uranium dioxide compacts prepared at a range of compaction pressures with and without wall lubricant. (From: The determination of density distributions in ceramic compacts using autoradiography, *Powder Technol.*, **16**, 107–122 (1977), reprinted by permission of Elsevier Sequoia S.A.)

For a general shape of mould, however, the determination of stress and density variations cannot be done without the help of a computer and a finite element code. It is the purpose of this section to introduce readers to the theoretical basis of continuum mechanics, and to give them some examples of the powerfulness of this method.

4.4.2 Continuum mechanics theory

A few basic definitions of continuum mechanics are given here in order to emphasize, in the following, the differences between ceramic powders and soils or metal powders (see also the Appendix for a description of tensor notations).

In the elastoplastic theory, the mechanical behaviour of a material may be summarized by two concepts: the **yield locus** and the **flow rule**.

Let us consider a small volume dV whose density is ρ_0, loaded with the stress σ. Starting from $\sigma = 0$ and loading monotonically, the element will first display elastic behaviour. Then, for a given σ_0 which depends on ρ_0, it will undergo some irreversible deformation. The yield locus $S(\rho_0)$ is the locus of these σ_0 in the stress space. In other words, it describes 'when' the powder yields.

The mechanical behaviour of a powder may depend not only on its density, but also on its microstructure (pore size distribution, anisotropy). The continuum mechanics theory is able to take it into account: for example, the yield locus might also depend on the 'fabric tensor' [58], or on other 'internal variables' describing the microstructure. The relative density is used here for the sake of simplicity.

Now the question is 'how' the powder yields, that is, what is its irreversible deformation (described by the plastic strain rate tensor $\dot{\varepsilon}_p$). This is the purpose of the flow rule, which gives the 'direction' and amplitude of the plastic strain rate tensor as a function of ρ_0 and σ_0.

The determination of the yield locus and flow rule would be impossible if all densities and all loading paths were to be examined. Luckily, owing to the isotropic properties of the material, the powder behaviour is determined only by the three stress 'invariants' p, q and K

$$p = \frac{\mathrm{tr}\,\sigma}{3},$$

$$q = (\bar{\sigma}_{ij}\bar{\sigma}_{ij})^{1/2}$$

where $\bar{\sigma}$ is the deviatoric part of σ,

$$K = \det(\bar{\sigma}).$$

It can be shown experimentally that the influence of K is usually negligible [59]. As a consequence, it is sufficient to determine, for each density ρ, the locus of the points (p, q) causing an irreversible deformation to the sample.

The functions p and q have rather clear physical significations: the first invariant is the mean pressure on the powder (under isostatic compaction, p is exactly the compaction pressure). The second gives an estimation of the 'distance' between the actual stress state and the nearer isostatic stress.

4.4.3 Experimental

For metal powders, four tests are used for the determination of f: tensile strength, free compression strength, uniaxial and isostatic compactions [60]. For a given density, each of these experiments gives one point in the (p, q) plane, and a curve (usually elliptic) is interpolated through these points (Fig. 4.20). The determination of the flow rule is straightforward since it is 'associated', i.e. the direction of the plastic strain rate is parallel to the gradient of the yield locus.

This is no longer true with ceramic powders. First, the positions of these four points in the stress space does not allow a precise interpolation: for a given density, the tensile and compressive strengths are too small compared to the compaction pressure (Fig. 4.20). Moreover, the flow rule of granular materials is not associated: the flow rule is not determined by the yield locus.

Owing to all these problems, it is necessary to use a device which is able to apply an axisymmetric stress on a sample, while measuring continuously its deformation. This is the purpose of the so called triaxial test (Fig. 4.21).

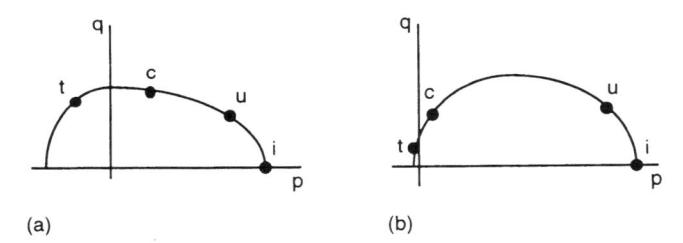

(a)   (b)

Fig. 4.20 Yield locus for: a = metal, b = ceramic powder. The points t, c, u, i represent respectively tensile strength, free compression strength, uniaxial compaction and hydrostatic compaction.

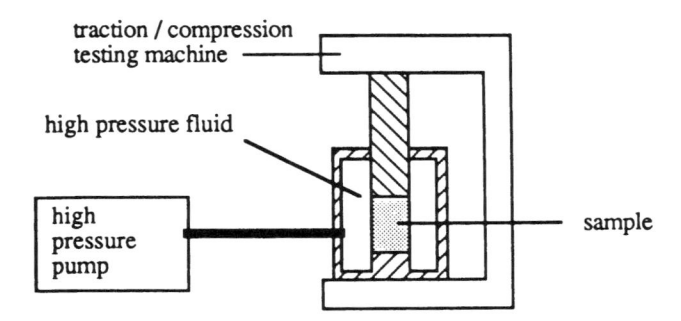

Fig. 4.21 Triaxial cell experiment.

In this test, a cylindrical sample is submitted to a radial pressure applied by a liquid, and to an axial pressure applied with a compression testing machine. It is prevented from fluid contact by a thin polymer jacket. Its axial deformation is measured by an extensometer, while its volumetric deformation is measured through either the fluid variations in the cell, the air pressure in the pores or the deformation of gauges placed on the sample.

The use of such devices is increasingly common in ceramic technology [4, 59, 61, 62].

4.4.4   Results

Table 4.2 shows that ceramic powders have some characteristics in common with soils, and others in common with metal powders. As a consequence, they will display a specific behaviour, using both soils and powder metallurgy concepts.

(a)   Elasticity constants

The microscopic friction law between spherical particles may be assumed to obey a Coulomb friction law, and the elastic repulsion a Hertzian law. From these assumptions, it can be demonstrated [63, 64] that the macroscopic elasticity is non-linear: shear and bulk moduli depend on the stress invariants.

The following results, from the author's experiments on a triaxial apparatus (zirconia powder), show that this is indeed the case (the influence of density has been shown to be negligible). From the same experiments, the Poisson's ratio was found to be constant.

The modulus of green compacts is rather low compared with that of the bulk material: this is one of the reasons for the important springback of ceramic powders upon unloading (Fig. 4.22).

Table 4.2 Comparison between different powders

	Particle behaviour	Packing	Deformation mode	Microscopic deformation mechanism
Metal powder	Plastic	Dense	Compaction	Plastic deformation
Sand	Rigid	Dense	Shear	Rearrangement of particles
Ceramic	Rigid	Loose	Compaction	Rearrangement of particles
Pharmaceutical	Brittle–plastic	Loose–dense	Compaction	Plastic deformation and fracture

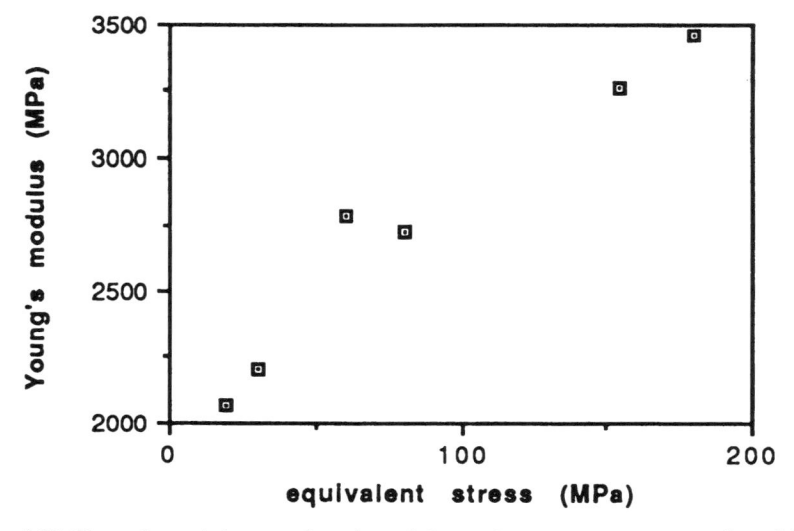

Fig. 4.22 Young's modulus as a function of the equivalent stress: $\sigma_{equ} = (p^2 + q^2)^{1/2}$.

(b)   Isostatic loading

The triaxial cell may be used to apply an isostatic load on the sample, while measuring continuously its vertical and radial dimensions. This curve is interesting on a 'mechanical' point of view, because it describes the density evolution along the p axis (the 'hardening law'). Figure 4.23 (from the same experiments on zirconia powders) shows that this curve is a straight line in a semi-logarithmic plot, as expected from the results described in section 4.3.

However, during the author's experiments along isostatic loadings, the axial deformation of the cylindrical sample was always smaller than the radial

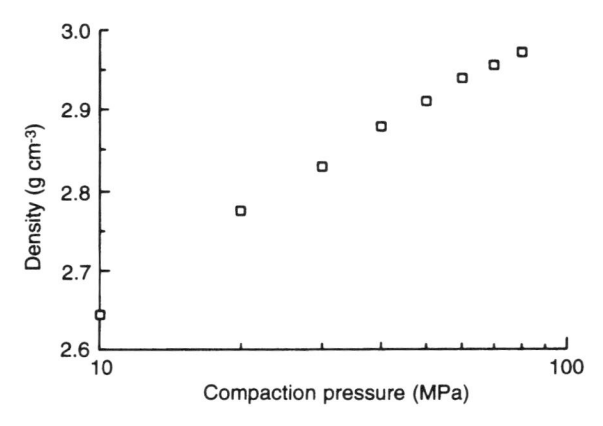

Fig. 4.23 Pressure/density relationship along a hydrostatic path.

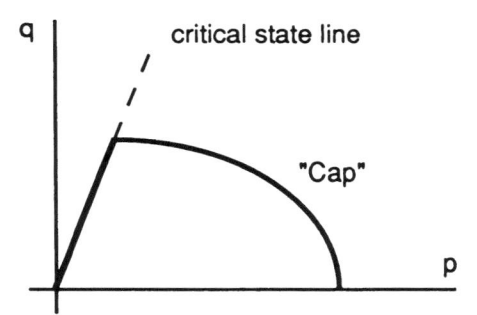

Fig. 4.24 Yield locus for ceramic powders.

deformation. The same phenomenon has been observed by other authors [61] and by Shima [59] who used a true triaxial cell. This result means that even a slight deviation (end effect, sample anisotropy) from a true isostatic load is able to cause a rather strong deviation from a pure isotropic deformation. Kolymbas [65] stated that this phenomenon is characteristic of highly compressible rigid granular materials. This might be important for the design of isostatic compaction moulds.

(c) Yield locus and flow rule

A complete description of the results is difficult since it involves several concepts from continuum mechanics. The main point is that the ceramic powder behaviour seems to be well reproduced by the so-called 'double hardening model', developed for soil mechanics by Vermeer [66]. The yield locus is defined by a straight line ('critical state line') and a linear or elliptical 'cap' (Fig. 4.24). The cap describes the compaction behaviour of the powder, while the critical state line is involved in high shear situations. The flow rule is a complicated function of stress and density.

4.4.5 Numerical simulation

The use of continuum mechanics is interesting for two reasons. The first is that is allows a better comparison between different powders than the classical tests. The second is that it allows the numerical calculation of stress and density variations in shaped moulds, thus allowing the engineer to forecast the problems that may arise during compaction.

The interested reader may find the basis for the understanding of numerical simulation in Gudehus [67]. Our purpose is to give some examples to demonstrate the usefulness of this method.

The first examples of such calculations are, to our knowledge, given by Strijbos [62]; other examples are described in references [4, 61, 68]. Broese van Groenou [68] reproduced satisfactorily its experimental density data for

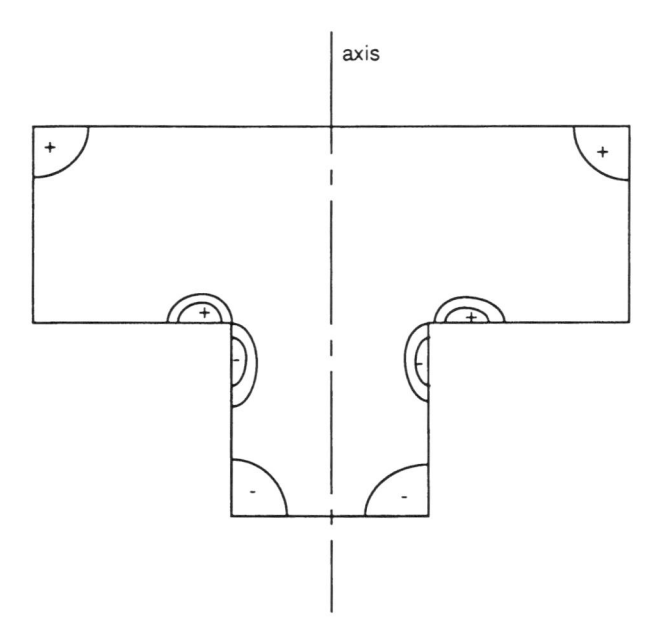

Fig. 4.25 Sketch of the density distribution in a mould with variable cross section.

a sample compacted with a hemispheric plunger. All these authors used roughly the same kind of model (double hardening).

During the compaction of a mould with variable cross section, calculations using a cap model with a straight cap [69] show that the density is higher slightly above the corner because of the high shear stress on this part of the mould. Under the corner, the density is lower because the high density region prevents the powder from flowing downwards.

Such a phenomenon has been proved experimentally by Morimoto [70] with a metallic powder. This author was also able to forecast a crack opening during densification, in the corner of the sample (Fig. 4.25).

The numerical simulation is also interesting for the forecast of tensile stresses upon unloading. Thompson [52] has shown that during unloading and ejection of a cylinder, the stress inhomogeneity during compaction turns into tensile stresses in the corner, resulting in the famous end capping problem. This author showed also that a slight ejection pressure over the sample is sufficient to overcome the tensile stresses until the green is completely ejected. This solution is widely used in industrial presses.

4.4.6   Conclusion

This section was a simple introduction to the theory of continuum mechanics. We tried to explain briefly its basis, and to show that it is able to illuminate some important features of the ceramic powder behaviour.

The main application of continuum mechanics is numerical simulation. We presented some examples published in the literature: a quantitative prediction of the experimental behaviour of the powder is indeed possible.

With the increasing power of computers, numerical simulation will certainly become an every day tool for the shape-forming engineer. However, there is still a great amount of work to be done in the experimental determination of the powder behaviour.

4.5 MICROSCOPIC BEHAVIOUR

So far we have considered the powder as a continuum, ignoring the internal structure of the sample. However, it is recognized that the pore size distribution, or even the pore morphology, will be of primary importance during sintering [21]. On the other hand, the green compact anisotropy induced by compaction will turn into differential shrinkage [71]. It may also influence the mechanical behaviour of the green compact, as is well known in soil mechanics [58]. These examples show the importance of microstructure evolution.

The first part of this section is dedicated to the study of packing evolution, i.e. porosity and anisotropy. Some results concerning changes in particle morphology and size will also be given.

On the other hand, an understanding of the behaviour of the individual particles during densification might help us to solve the macroscopic problems. Considering the relatively simple structure of a particle packing, it has long been thought that the calculation of the macroscopic behaviour of the powder might be possible from the knowledge of the microscopic interaction laws. The second part of this section is dedicated to this important question.

4.5.1 Microstructure evolution

(a) Experimental

Compared to other people working with granular materials, ceramists are unlucky because of the very small size of their powders. This makes direct observation of the packing evolution during compaction difficult: such beautiful experiments as those of Oda [72] or Fishmeister [73], who studied directly the packing of sand or metal powder, are very difficult to perform. It is, of course, possible to examine fracture surfaces or polished surfaces (the specimen being filled with resin). But these observations are not precise: the behaviour of the individual particle is difficult to examine.

However, a lot of indirect methods are available for microstructural examination. The most well known is probably mercury porosimetry. The interpretation of its results is to be conducted carefully: it is well known that this method measures the intrusion diameter of the pores, rather than their true diameter. This must be kept in mind when examining special structures [74], such as samples obtained with a platelet-like powder.

The measurement of adsorption isotherms (water, nitrogen) is also interesting. The most simple method gives the particle surface area, but the adsorption isotherms may give far more information, including pore size distribution and surface potentials. Stanley-Wood used this method extensively [75].

Among the non-destructive testing methods, NMR has a great potential. It gives a direct image of the macropores in a geen compact filled with water, up to a few hundreds of microns in size. This is particularly interesting for wet processing, since the water necessary for the measurement is 'naturally' present in the sample [76]. The ultrasonic wave velocity is also of interest, and may supply a lot of information [77].

(b)   Results

The **pore size distribution** obtained after dry compaction of a monosized powder is rather monomodal, with sometimes a tail towards the large pore size. As the compaction pressure increases, the mean size of the pores decreases and the width of the distribution increases [10]. However, the larger pores seem to resist the high pressure [36], or may even re-appear at high pressure due to the springback [78]. Naito showed that the destruction of the larger pores is better when the stress relaxation of the packing is improved with a lubricant.

It is interesting to note that these effects are rather different in pressure filtration: in this case, the width of the pore size distribution seems to decrease as the pressure increases. For a given density, the mean pore size is always smaller in the wet case than in the dry case [79].

The most important powder characteristic, considering the pore size distribution, seems to be the occurrence of aggregates. When the powder contains enough aggregates, the pore size distribution becomes bimodal [22]. Upon loading, the size and the number of these large pores decrease, while only the number of the small pores decreases [80].

The particle morphology is also important for the pore size distribution. Yamagushi *et al.* [21] showed that the pores of a platelet-like powder are more uniform and smaller than those of a spherical powder (for a given green density). The pore morphology will certainly also be different.

Considering the hardness of ceramic particles, the surface involved in contact between particles is likely to be very small compared to the specific surface of the powder. However, there is some evidence that the **specific surface** of the green compact decreases during compaction [81]. This phenomenon, however, is still controversial.

Due to the particle hardness, too, **fracture of primary particles** has been scarcely mentioned (while aggregate fracture is very common). It happens only for very elongated particles, which may be submitted to bending stresses [20]. On the contrary, the size evolution of pharmaceutical particles during compaction is a common phenomenon [23, 24].

Turning now to the **anisotropy** of the packing, one must note that there are two kinds of anisotropy. The first arises from the particles themselves, if they are not spherical; for example, after compaction platelets are perpendicular to the loading axis [21]. The second arises from the packing itself: the directions of the particle/particle bonds are not distributed randomly.

As usual in granular media mechanics, the soil engineers pioneered the measurement of packing anisotropy. The best way to characterize it is the so-called 'fabric tensor' [82] which describes the proportion of bonds oriented towards a given direction. The evolution of the fabric tensor under stress or deformation has been studied for cohesionless powders or barrels [82, 83]. The main conclusion is that under loading, the packing moves so that the contacts normals are oriented mainly in the loading direction.

This effect may be verified in ceramics by comparing the shrinkage in the axial and radial directions of a green body submitted to uniaxial compaction. The author's experiments show that the radial shrinkage is always greater than the axial. Fang and Hsie [71] found that the relative magnitude of the axial and radial shrinkages depends on the shape forming process. Under dry pressing conditions, the shrinkage is greater in the radial direction. On the contrary, the axial shrinkage is greater than the radial shrinkage for pressure filtrated compacts.

4.5.2 Micromechanics of compaction

This part of the microstructure study is dedicated to a more fundamental approach: is it possible to determine theoretically the density and compaction response of a packing of particles whose geometry and interaction laws are known? The first step of this approach is to study the 'static' packing: what is the density and structure of a packing of particles whose characteristics are known? The second step is to determine the stress/strain relationship of this packing. Both have not yet received definite answers. An extensive review of this question is given by Bortzmeyer [4].

(a) Statics of packing

The question 'what will be the density of a packing of particles whose characteristics are known' has received a lot of attention. As usual, the earliest models have been derived for soils mechanics, so that they deal only with cohesionless particles.

To our knowledge, no rigorous theoretical answer to this question has been given. A few authors tried to use the formalism of statistical thermodynamics (maximization of the 'entropy' of the packing). However, it is difficult to define rigorously this entropy. A criticism of most of these models is given by Kanatani [14].

A more practical approach is to start from the result concerning a mono-sized population (relative density 0.63) and to calculate the density of

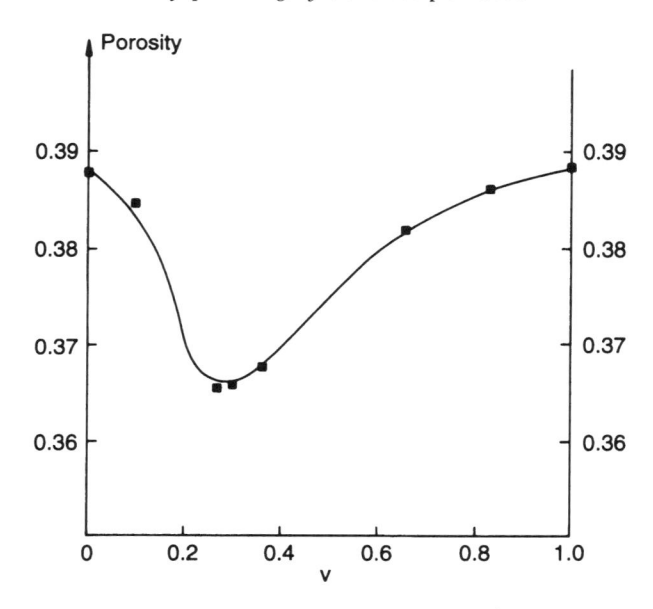

Fig. 4.26 Variation of the porosity of a binary mixture ($d^2/d_1 = 4$) with the volume fraction of small spheres. (From: [86] Properties of disordered sphere packings. I. Geometric structure: statistical model numerical simulations and experimental results, *Powder Technol.*, **46**, 121–131 (1986), reprinted by permission of Elsevier Sequoia SA.)

a multimodal population with the help of geometrical functions representing the steric interaction between particles of different sizes [84, 85]. For example, Fig. 4.26 shows the density as a function of the volume fraction of small spheres for a bimodal population: mixing two sizes always results in an increase in density [86].

On the basis of such a geometrical model, Ouchiyama and Tanaka [87] investigated the influence of particle cohesion on the relative density. However, these theoretical methods are restricted to simple interaction laws. A more general study, involving for example the vaults effects resulting from the mass movement of the particles, cannot be achieved without the help of a computer [88], which is why the numerical modelization of packing construction takes an increasing place in the literature [89, 90].

(b) Dynamics of the packing

Since the prediction of the packing static density is still controversial, it is not surprising that a complete calculation of the packing dynamic behaviour is still lacking. A complete survey of this interesting topic is far beyond the scope of this chapter [4]. It is worth explaining, however, what makes this calculation so difficult.

The main problem is that forces and movements are not homogeneous among the particles [91, 92]: forces are transmitted in the packing through several 'chains' of stressed particles; deformation obeys the so-called 'Horne law' (the packing deformation is the result of the relative movements of large rigid blocks, rolling and sliding one over the other, and recombining during the deformation).

This situation is very difficult to model theoretically, but if one tries to 'forget' this inhomogeneity and to calculate the behaviour of the packing as if the forces were homogeneous, a paradoxal situation arises: under an isostatic loading, the force on each bond is normal to the bond tangent plane. As a consequence, the contact cannot slip and no deformation occurs. Thus, it is not possible to densify a powder with an isostatic loading.

(c)   Numerical simulation

Since the 'manual' calculation does not work, the best solution is to devise a micromechanical model which simulates on a computer the behaviour of an assembly of rigid discs (Fig. 4.27). In this kind of calculation, the interparticle forces are known (van der Waals, elastic repulsion, sliding friction) and the computer determines the behaviour of the whole packing in a quasistatic deformation. This kind of simulation is now well known in the case of soils mechanics; some three dimensional applications have begun to appear [93].

The following figures come from the author's thesis. They describe the results of a numerical simulation of a collection of discs. The interaction forces are a van der Waals attraction, a linear-elastic repulsion, a Coulomb friction and a rolling friction.

Figure 4.28 shows the density/pressure relationship of this material at the beginning of an uniaxial compaction. It is seen that, although the walls move smoothly with time, the wall pressure displays strong instabilities.

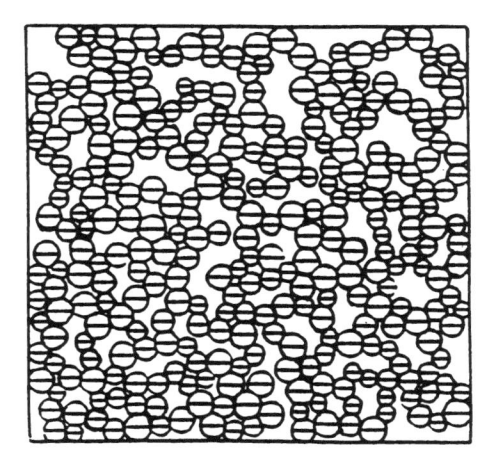

Fig. 4.27 Two dimensional packing used in computer simulation of ceramic powder compaction.

Dry pressing of ceramic powders

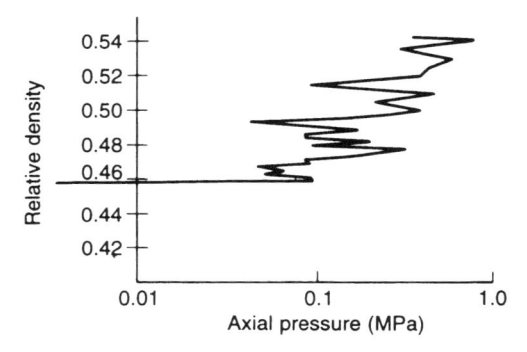

Fig. 4.28 Density/pressure relationship at the beginning of the densification showing strong instabilities.

The mechanism of these instabilities may be explained in the following way: at the beginning of the densification, the discs roll one over the other rather easily; the compaction pressure is small. After a while, they reach a blocking configuration which can be compared to a vault or a buttress. This structure resists the deformation, so the pressure rises rapidly. When it is high enough, the structure collapses (buckling), the pressure decreases strongly and the same phenomenon starts again. This means that, at least in two dimensions, compaction is an unstable process on a micromechanical scale. Kuhn *et al.* [3] reached the same conclusion with experiments on cylinders.

These instabilities are difficult to manage in a computer program (convergence difficulties): 'soft' interaction laws were used to reduce them. As a consequence, the numerical values of the compaction pressure are small. Figures 4.29 and 4.30 show, however, that the compaction behaviour of the ceramic powder is qualitatively reproduced.

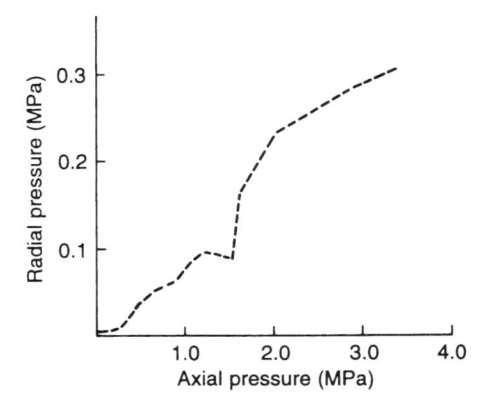

Fig. 4.29 Radial pressure against axial pressure during uniaxial compaction of the two dimensional simulation.

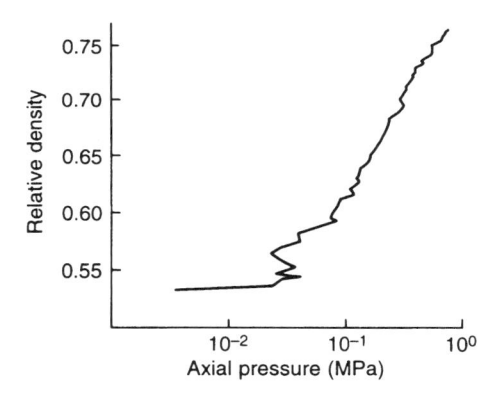

Fig. 4.30 Density/pressure relationship during uniaxial compaction of the two dimensional simulation.

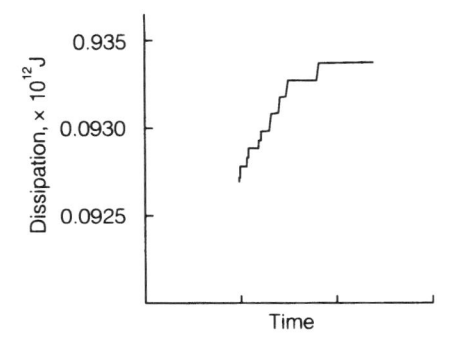

Fig. 4.31 Energy dissipation upon unloading.

We mentioned earlier (section 4.3) that the unloading is not purely elastic; even this small phenomenon is reproduced by our simulation (Fig. 4.31).

Figure 4.32 showing a packing loaded isostatically, sheds light on the problem of force inhomogeneity described earlier. The interparticle forces are represented as bold lines: the thicker the line, the higher the force. It is seen that the pressure is transmitted from one side to another by a few 'chains' of particles.

4.5.3 Conclusion

This section was dedicated to the study of packing microstructure. Experiments on 'real' samples with mercury porosimetry, Brunauer–Emmelt–Teller test, etc. allow a precise description of the structure evolution. Classical results (pore size distribution) as well as less classical (anisotropy) were described.

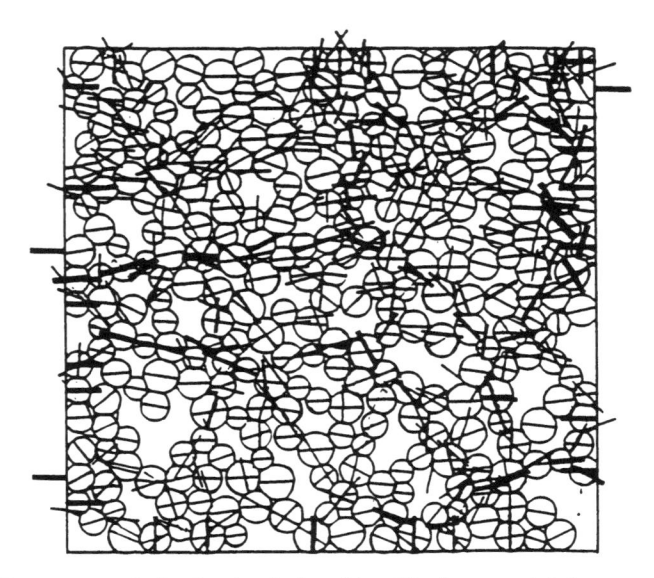

Fig. 4.32 Force transmission in a loaded packing. The interparticle forces are represented by bold lines: the thicker the line, the greater the force.

This is important since the sintering step is very dependent on sample structure.

On the other hand, experiments on model systems (cylinders or numerical simulation of discs) gave us an interesting insight into the microscopic behaviour of the powder. The micromechanism of compaction was seen to be the buckling of the particle structure.

4.6 CONCLUSIONS

This brief review of current research on dry pressing was intended to examine three basic approaches to the compaction process.

First, the 'classical' approach: pressure/density relationship, radial pressure coefficient, wall friction coefficient and green tensile strength. These four experiments form a good basis for a rich comparison of powders. The influence of powder characteristics and process parameters on their results is quite well known.

Second, the 'mechanical' approach: the classical method cannot predict the stress and density variations in shaped moulds; the use of continuum mechanics and finite element calculation, is necessary. This field is rapidly developing in ceramic technology.

Third, the 'microscopic' approach: the study of the green microstructure has shed light on several important parameters including pore size distribution and anisotropy. On the other hand, the numerical study of the packing behaviour has a great potential for a fundamental understanding of dry pressing.

We hope that the reader has been convinced that a simultaneous development of these three approaches is necessary for a complete description of the compaction process.

APPENDIX: TENSOR NOTATION FOR STRESS

The stress on a volume element dV of the powder is defined by a symmetrical tensor σ

$$\sigma = \begin{pmatrix} \sigma_{xx} & \tau_{xy} & \tau_{xz} \\ \tau_{xy} & \sigma_{yy} & \tau_{yz} \\ \tau_{xz} & \tau_{yz} & \sigma_{zz} \end{pmatrix}$$

The significance of the stresses σ_{xx}, τ_{xy} and τ_{xz} is given in Fig. 4.A1

The powder behaviour is supposed isotropic. As a consequence, it depends only on the three components of σ that are invariant upon an axis rotation. These are its trace p, the norm of its deviatoric part q, and its determinant. Usually, the influence of the third stress invariant is assumed negligible and the behaviour depends only on

$$p = -\frac{\text{tr}(\sigma)}{3} \qquad \text{first stress invariant}$$

$$q = (\bar{\sigma}_{ij}\bar{\sigma}_{ij})^{1/2} \qquad \text{second stress invariant}$$

where $\bar{\sigma} = \sigma - pl$, l being the unity tensor.

The physical signification of p and q is clear: the first invariant is the mean pressure on the material (for an isostatic compaction, p is equal to the compaction pressure). The second gives an indication of the 'distance' between the stress state and the nearest isostatic stress.

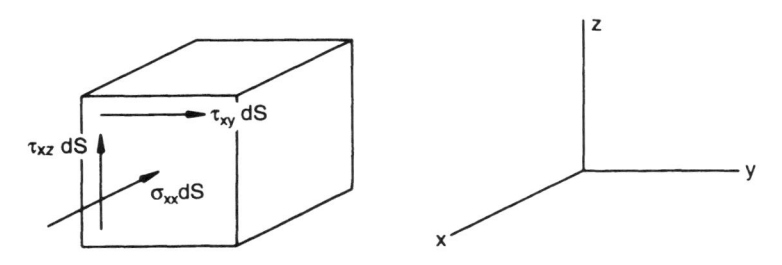

Fig. 4.A1 Notation for the partial stresses: the force on the face dS normal to the $\vec{x}$ axis is $dS\,(\sigma_{xx}\vec{x} + \tau_{xy}\vec{y} + \vec{\tau}_{xz}z)$.

REFERENCES

1. Quinn, D.B., Bedford, R.E. and Kennard, F.L. (1984) Dry-bag isostatic pressing and contour grinding of technical ceramics, in *Advances in Ceramics, Vol. 9, Forming of Ceramics* (eds J.A. Mangels and G.L. Messing), The American Ceramic Society, Columbus, pp. 4–15.
2. Mac Entire, B.J. (1984) Tooling design for wet-bag isostatic pressing, in *Advances in Ceramics, Vol. 9, Forming of Ceramics* (eds J.A. Mangels and G.L. Messing), The American Ceramics Society, Columbus, pp. 16–31.
3. Kuhn, L., Mac Meeking, R.M. and Lange, F.F. (1989) in *Powders and Grains, Proceedings of the 1st International Conference on Micromechanics of Granular Materials*, Clermont-Ferrand, France, 4–8 September (eds J. Biarez and E. Gourves), Balkema, Rotterdam, pp. 331–8.
4. Bortzmeyer, D. (1990) Compaction of ceramic powders. PhD Thesis, Ecole des Mines de Paris.
5. Matsumoto, R.L.K. (1986) Generation of powder compaction response diagrams, *J. Am. Ceram. Soc.*, **69**(10), C246–C247.
6. Lukasiewicz, S.J. and Reed, J.S. (1978) Character and compaction response of spray dried agglomerates, *Am. Ceram. Soc. Bull.*, **57**(9), 798–801.
7. Groot Zevert, W.F.M., Winnubst, A.J.A., Theunissen, G.S.A.M. and Burggraaf, A.J. (1990) Powder preparation and compaction behaviour of fine-grained Y–TZP, *J. Mater. Sci.*, **25**, 3449–55.
8. Harvey, J.W. and Johnson, D.W. Binder (1980) Systems in ferrite, *Am. Ceram. Soc. Bull.*, **59**(6), 637–45.
9. Brewer, J.A., Moore, R.H., Reed, J.S. (1981) Effect of relative humidity on the compaction of barium titanate and manganese zinc ferrite agglomerates containing polyvinyl alcohol, *Am. Ceram. Soc. Bull.*, **60**(2), 212.
10. Frey, R.G. and Halloran, J.W. (1984) Compaction behaviour of spray dried alumina, *J. Am. Ceram. Soc.*, **67**(3), 199–203.
11. Di Milia R.A. and Reed, J.S. (1983) Dependence of compaction of the glass transition temperature of the binder phase, *Am. Ceram. Soc. Bull.*, **62**(4), 484–8.
12. Nies, C.W. and Messing, G.L. (1984) Binder hardness and plasticity in granule compaction, in *Advances in Ceramics, Vol 9, Forming of Ceramics*, (eds J.A. Mangels and G.L. Messing), The American Ceramic Society, pp. 58–66.
13. Caligaris, R.E., Topolewsky, R., Maggi, P. and Brog, F. (1985) Compaction behavior of ceramic powders, *Powder Technol.*, **42**, 263–7.
14. Kanatani, K.I. (1981) The use of entropy in the description of granular materials, *Powder Technol.*, **30**, 217–23.
15. Kawakita, K. and Lüdde, K.H. (1970) Some considerations on powder compression equations, *Powder Technol.*, **4**, 61–8.
16. Gonthier, Y. (1984) Contribution à l'etude du comportement mécanique des poudres pharmaceutiques sous pression, Thesis, University of Grenoble.
17. Chaklader, A.C.D. and Bhattacharya, S.K. (1987) Effect of additives on the cold compaction behaviour of SiC powders, in *Sintering 85*, (eds G.C. Kuczynski, D.P. Uskokovic, H. Palmour and M.M. Ristic) Plenum Press, pp. 359–70.
18. Carless, J.E. and Leigh, S (1974) Compression characteristics of powders: radial die wall pressure transmission and density changes, *J. Pharm. Pharma.*, **26**, 289–97.
19. Zheng, J. and Reed, J.S. (1988) Particle and granule parameters affecting compaction efficiency in dry pressing, *J. Am. Ceram. Soc.*, **71**(11), C 456–C458.
20. Leiser, D.B., Whittemore, O.J. (1970) Compaction behavior of ceramic particles, *Am. Ceram. Soc. Bull.*, **49**(8), 714–717.
21. Yamaguchi, T. and Kosha, H (1981) Sintering of acicular Fe_2O_3 powders, *J. Am. Ceram. Soc.*, **64**(5) C84–C85.

22. Ciftcioglu, M., Akinc, M. and Burkhart, L. (1987) Effect of agglomerate strength on sintered density of yttria powders containing agglomerates of monosized spheres, *J. Am. Ceram. Soc.*, **70**(11), C329–C334.
23. Carless, J.E. and Sheak, A. (1976) Changes in the particle size distribution during tableting of sulphathiazole powder, *J. Pharm. Pharmac.*, **28**, 17–22.
24. Butler, P.B., Haworth, M.E., Elban, W.L. and Coyne, P.J. (1990) Particle morphology characterization of quasi-statistically compacted sucrose, *Powder Technol.*, **62**, 171–81.
25. Masters, K. (1979) *Spray Drying Handbook*, 4th edn, Longman Scientific and Technical (J. Wiley copublisher), New York.
26. Lukasiewicz, S.J. (1989) Spray drying ceramic powders, *J. Am. Ceram. Soc.*, **72**(4), 617–24.
27. Liang, B. and King, C.J. (1991) Factors influencing flow patterns, temperature fields and consequent drying rates in spray drying, *Drying Technol.*, **9**(1), 1–25.
28. Bast, R. (1990) Organic additives for dry pressing, *Interceram.*, **39**(6), 13–14.
29. Kuno, H. and Okada, J. (1982) The compaction process and deformability of granules, *Powder Technol.*, **33**, 73–9.
30. Oberacker, R., Ottenstein, A. and Thümmler, F. (1988) Characterisation of granules by measurement of load–deformation curves with a newly developed strength tester, in *Proceedings of the 2nd International Conference on Ceramic Powder Processing Science*, 12–14 October, Berchtesgaden, Germany, pp. 37–47.
31. van der Zwan J. (1989) Granule strength and compaction behaviour of agglomerated materials in *Euro–Ceramics, Vol. 1, Processing of Ceramics* (eds G.de With, R.A. Terpstra and R. Metselaar), Elsevier, London, pp. 1238–42.
32. DiMilia R.A. and Reed J.S. (1983) Stress transmission during the compaction of a spray-dried alumina powder in a steel die, *J. Am. Ceram. Soc.*, **66**(9), 667–72.
33. Uetsamu, K., Kim, J.-Y., Miyashita, M., Uchida, N. and Saito, K (1990) Direct observation of internal structure in spray-dried alumina, *J. Am. Ceram. Soc.*, **73**(8), 2555–7.
34. Youshaw, R.A., Halloran, J.W. (1982) Compaction of spray dried powders, *Am. Ceram. Soc. Bull.*, **61**(2), 227–230.
35. Messing, G.L., Markhoff, C.S. and Mac Coy, L.G. (1982) Characterization of ceramic powder compaction, *Am. Ceram. Soc. Bull.*, **61**(8), 857–60.
36. Naito, N. (1979) Pore size distribution during compaction and early stage sintering of Si_3N_4, MS Thesis, Lawrence Berkeley Laboratory University of California.
37. Matsuo, Y., Nishimura, T., Jinbo, K., Yasuda, K. and Kimura, S. (1987) Development of cyclic-CIP and its application to powder forming, *Yogyo-Kyokai-Shi* **95**(12), 1226–31.
38. Nishimura, T., Jinbo, K., Matsuo, Y. and Kimura, S. (1990) Forming of ceramic powders by cyclic-CIP. Effect of bias pressure, *J. Ceram. Soc. Jap. Intn. Edn*, **98**, 742–5.
39. Emeruwa, E., Jarrige, J. and Mexmain, J. (1989) Ferrite powder compaction with ultrasonic assistance, in *Euro Ceramics, Vol. 1, Processing of Ceramics* (eds G. de With, R.A. Terpstra and R. Metselaar), Elsevier, London, pp. 1248–52.
40. Strijbos, S., Rankin, P.J., Klein Wassink, R.J., Bannink, J. and Oudemans, G.J. (1977) Stresses occurring during one sided die compaction of powders, *Powder Technol.* **18**, 187–200.
41. Henke, M., Klemm, U. and Sobek, D. (1986) Determination of specific parameters in dry pressing of ceramic powders, *J. Powder Bulk. Solids Technol.*, **10**(1), 9–14.
42. Strijbos, S. and Knaapen, A.C. (1977) Mechanical properties of a ferrite powder and its granulate; *Sci. Ceram.*, **9**, 477–485.
43. Kendall, K. (1986) Inadequacy of Coulomb's friction law for particle assemblies, *Letters to Nature*, **319**, 203–5.

44. Claussen, N. and Jahn, J. (1970) Green strength of metal and ceramic compacts as determined by the indirect tensile test, *Powder Metall. Int.* **2**(3) 87–90.
45. Marion, R.K. and Johnstone, J.K. (1977) Parametric study of the diametral compression test for ceramics *Am. Ceram. Soc. Bull.*, **56**(11), 958–1002.
46. Mehrabadi, M.M. and Nemat–Nasser, S. (1982) On statistical description of stress and fabric in granular materials, *Int. J. Num. Anal. Meth. Geomechanics*, **6**, 95–108.
47. Schubert, H. (1975) Tensile strength of agglomerates, *Powder Technol.*, **11**, 107–119.
48. Kendall, K., Mac Alford, N. and Birchall, J.D. (1986) The strength of green bodies, *Special Ceramics* **8**, 255–65.
49. Adams, M.J., Williams, D. and Williams, J.G. (1989) The use of linear elastic fracture mechanics for particulate solids, *J. Mater. Sci.*, **24**, 1772–16.
50. Bortzmeyer, D. (1992) Tensile strength of ceramic powders. *J. Mater. Sci.*, **27**, 3305–8.
51. Thompson, R.A. (1981) Mechanics of powder pressing. I. Model for powder densification, *Am. Ceram. Soc. Bull.*, **60**(2), 237–43.
52. Thompson, R.A. (1981) Mechanics of powder pressing. II. Finite element analysis of end-capping in pressed green powders, *Am. Ceram. Soc. Bull.*, **60**(2), 244–7.
53. Thompson, R.A. (1981) Mechanics of powder pressing. III. Model for the green strength of pressed powders, *Am. Ceram. Soc. Bull.*, **60**(2), 248–51.
54. Broese van Groenou, A. (1978) Pressing of ceramic powders: a review of recent work, *Powder Met. Int.*, **10**(4), 206–11
55. Broese van Groenou, A. and Knaapen (1980) Density variations in die compacted powders, in *Science of Ceramics*, Proceedings of the 10th International Conference on Science of Ceramics, Berchtesgaden, 1979, Vol. 10 (ed H. Hausner), Deutsche Keramische Gesellschaft pp.100–105.
56. Ellington, W.A., Ackerman, J.L., Garrido, L., Weyand, J.D. and Di Milia, R.A. (1987) Characterization of porosity in green state and partially densified Al_2O_3 by nuclear magnetic resonance imaging, *Ceram. Eng. Sci. Proc.*, **8**(7–8), 503–12.
57. Mac Leod, H.M. and Marshall, K. (1977) The determination of density distribution in ceramic compacts using autoradiography, *Powder Technol.*, **16**, 107–22.
58. Oda, M. and Sudoo, T. (1989) Fabric tensor showing anisotropy of granular soils and its application to soil plasticity, in, *Powders and Grains, Proceedings of the 1st International Conference on Micromechanics of Granular Media*, Clermont-Ferrand, France, 4–9 September 1989 (eds J. Biarez and R. Gourvès), Balkema, Rotterdam., pp. 155– 162.
59. Shima, S. and Mimura, K. (1986) Densification behavior of ceramic powders, *Int. J Mech. Sci.*, **28**(1), 53–9.
60. Abouaf, M. (1985) Modélisation de la compaction de poudres métalliques frittées. Doctoral Thesis, University of Grenoble.
61. Hehenberger, M., Samuelson, P., Alm, O., Nilsson, L. and Olofsson, T. (1982) Experimental and theoretical study of powder compaction, in *Proc. IUTAM Conf. on Deformation and Failure of Granular Material*, Delft, 31 August-3 September, pp. 381–90.
62. Strijbos, S. and Vermeer P.A. (1978) Stress and density distributions in the compaction of powders, in *Processing of Crystalline Ceramics* (eds H. Palmour, R.F. Davis and T. M. Hare), Mater Sci. Res., **11**, 113–123.
63. Bathurst, R.J. and Rothenburg L. (1988) Micromechanical aspects of isotropic granular assemblies with linear contact interactions, *J. Appl. Mech.*, **55**, 17–23.
64. Jenkins, J.T., Cundall, P.A. and Ishibashi, I. (1989) in *Powders and Grains, Proceedings of the 1st International Conference on Micromechanics of Granular Materials*, Clermont-Ferrand, France, 4–8 September 1989 (eds J. Biarez and R. Gourvès), Balkema Rotterdam, pp. 257–264.

65. Kolymbas, D. and Wu, W. (1990) Recent results of triaxial tests with granular materials. *Powder Technol.*, **60**, 99–119.
66. Vermeer, P.A. (1977) A double hardening model for sand, *Delft Progress Report Civil Engineering*, **2**, 303–20.
67. Gudehus, G. (1977) *Finite Elements in Geomechanics*, (ed. G. Gudehus) John Wiley and Sons, NY.
68. Broese van Groenou, A. (1982) Theory of dust pressing, *Interceram* **31**(6), 1–10.
69. Bortzmeyer, D. (1992) Modelling ceramic powder compaction. *Powder Technol*, **70**(2), 131–9.
70. Morimoto, Y., Hayashi T. and Takei, T. (1982) Mechanical behavior of powders during compaction in a mould with variable cross sections, *Int. J. Powder Metall. Powder Technol.*, **18**(2), 129–145.
71. Fang, T.-T. and Hsieh H.-L. (1988) Effects of pressing methods on the sintering behavior of high purity $BaTiO_3$, *J. Mater. Sci. Lett.*, **7**, 187–8.
72. Oda, M (1977) Coordination number and its relation to shear strength of granular material, *Soils Found.*, **17**(2), 29–42.
73. Fischmeister, H.F., Arzt, E. and Olsson, L.R. (1978) Particule deformation and sliding during compaction of spherical powders: a study by quantitative metallography, *Powder Metall.*, **21**(4), 179–185.
74. Lee, H.H.D. (1990) Validity of using mercury porosimetry to characterize and pore structure of ceramic green compacts, *J. Am. Ceram. Soc.*, **73**(8), 2309–15.
75. Stanley-Wood, N.G., Abdelkarim, A., Johansson, M.E., Sadeghnejad, G. and Osborne, N. (1990) The variation in, and correlation of, the energetic potential and surface areas of powders with degree of uniaxial compaction stress, *Powder Technol.*, **60**, 15–26.
76. Karunanithy, S. and Mooibroek, S. (1989) Detection of physical flaws in alumina reinforced with SiC fibres by imaging in the green state, *J. Mater. Sci.*, **24**, 3686–90.
77. Kupperman, D.S. and Karplus, H.B. (1984) Ultrasonic wave propagation characteristics of green ceramics, *Ceram. Bull.*, **63**(12), 1505–9.
78. Hsieh, H.-L. and Fang, T.-T. (1989) Effects of powder processing on the green compacts of high-purity $BaTiO_3$, *J. Am. Ceram. Soc.*, **72**(1), 142–5.
79. Hsieh, H.-L. and Fang, T.-T. (1990) Effect of green states on sintering behavior and microstructural evolution of high-purity barium titanate, *J. Am. Ceram. Soc.*, **73**(6), 1566–73.
80. Dynys, F.W. and Halloran, J.W. (1983) Compaction of aggregated alumina powders, *J. Am. Ceram. Soc.*, **66**(9), 655–9.
81. Stanley-Wood, N.G., Sarrafi, M. and Lagarde, S. (1988) Characterisation of uniaxially compacted titanium dioxide in the stress range 125–20 000 kPa by water adsorption, in *Proceedings of the 13th Annual Powder and Bulk Solids Conference* 9–2 May 1988, Rosemont, Illinois, pp. 471–82.
82. Oda, M. (1978) Significance of fabric in granular material, in *Proceedings of a US–Japan Seminar on Continuum Mechanics and Statistical Approaches in the Mechanics of Granular Materials* Sendai, Japan, 5–9 June 1978 (eds S.C. Cowin and M. Satake), Gakujutsu Bunken Fukyu-Kai, Tokyo, Japan, pp. 7–26.
83. Umeya, K., Hara, R. and Kikuta, J. (1975) On two dimensional shear test by model powders, *J. Chem. Eng. Japan*, **8**(1), 56–62.
84. Yu, A.B. and Standish, N. (1988) An analytical-parametric theory of the random packing of particles, *Powder Technol.*, **55**, 171–86.
85. Stovall, T., De Larrard, F. and Buil, M. (1986) Linear packing density model of grain mixtures, *Powder Technol.*, **48**, 1–12.
86. Oger, L., Troadec, J.P., Bideau, D., Dodds, J.A. and Powell, M.J. (1986) Properties of disordered sphere packings. I. Geometric structure: statistical model, numerical simulation and experimental results, *Powder Technol.*, **46**, 121–31.

87. Ouchiyama, N. and Tanaka, T. (1986) Porosity estimation from particle size distribution, *Ind. Eng. Chem. Fundam.*, **25**, 125–9.
88. Thomas, G., Missiaen, J.M. and Rouille, L. (1989) in, *Powders and Grains, Proceedings of the 1st International Conference on Micromechanics of Granular Media*, Clermont-Ferrand, France, 4–8 September 1989 (eds J. Biarez and E. Gourves), Balkema, Rotterdam, pp. 99–104.
89. Soppe, W. (1990) Computer simulation of random packing of hard spheres, *Powder Technol.*, **62**, 189–96.
90. Rodriguez, J., Allibert, C.H. and Chaix, J.M. (1986) A computer method for random packing of spheres of unequal size, *Powder Technol.*, **47**, 25–33.
91. Oda, M. and Konishi, J. (1974) Microscopic deformation mechanism of granular material in simple shear, *Soils and Foundations*, **14**(4), 25–38.
92. Konishi, J. (1978) Microscopic model studies on the mechanical behaviour of granular materials, in *US-japan Seminar on Continuum Mechanics and Statistical Approaches in the Mechanics of Granular Materials*, Sendai, Japan, 5–9 June 1978 (eds S.C. Cowin and M. Satake), Proceedings of a Gakujutsu Bunken Fukyu-Kai, Tokyo, Japan, pp. 27–45.
93. Cundall, P.A., Jenkins, J.T. and Ishibashi, I. (1989) in *Powders and Grains, Proceedings of the 1st International Conference on Micromechanics of Granular Materials*. Clermont-Ferrand France, 4–8 September 1989 (eds J. Biarez and E. Gourves), Balkema Rotterdam, pp. 319–22.
94. Niesz, D.E. and Bennett, R.B. (1972) Strength characterization of powder aggregates, *Am. Ceram. Soc. Bull.*, **51**(9), 677–80.

CHAPTER 5

The principles of tape casting and tape casting applications

R.E. Mistler

5.1 INTRODUCTION

Although the title of this chapter is 'The principles of tape casting and tape casting applications', it is important to provide background information on other competing processing technologies which are currently used to form thin sheets of ceramic materials. These will be discussed in the introduction, since the bulk of the chapter will be devoted to the subject of tape casting.

Historically, small-area substrates have been fabricated using dry pressing or extrusion techniques. Both of these technologies have been described in detail in other chapters of this book. As the demand for larger surface area and thinner cross section substrates increased, two novel fabrication processes were developed by the ceramic industry. One of these, tape casting, is the subject for discussion in this chapter. The other, roll compaction or roll forming, has not received as much attention in the literature since it is predominantly a process which has been developed by companies for use as a proprietary fabrication technique for a product of commerce which they are selling.

Roll compaction is a process which was first introduced by Gladding, McBean & Co. in the USA in the early 1950s. At least two patents were issued as a result of this seminal work [1, 2]. As the patents so eloquently state 'Definite limitations as to minimum thickness of tile so pressed (by dry pressing) have been found to exist and it is virtually impossible to compress and process normal bodies of finished tile with a shape and size where the width or face dimension of the tile is more than twenty times the thickness of the tile. Ordinary methods of compression such as are used in tile presses (including single-acting and double-acting types of presses) cannot be employed in the preparation of extremely thin wafers of ceramic material for use

Ceramic Processing. Edited by R.A. Terpstra, P.P.A.C. Pex and A.H. de Vries.
Published in 1995 by Chapman & Hall, London. ISBN 0 412 59830 2

as separators, insulators, etc.' [1]. The patent goes on to state that the invention of roll compaction permits the manufacture of thin sheets of ceramic which have face dimensions or widths from 30 to 3000 times the thickness, and actual sizes for width and length which extend to metres. The process of roll compaction is very similar to that for dry pressing. The powder preparation techniques, with the exception of the organic systems utilized are almost identical. Figure 5.1 is a schematic representation of the powder preparation portion of the process for the production of alumina substrates for thick film applications. This procedure is identical to that used in the preparation of spray-dried powders for many dry pressing operations. The major difference is in the type and quantity of binder and plasticizer which are used. Usually for roll compaction water-based acrylics, wax emulsions, or polyvinyl butyrals are used as binders with plasticizers added to provide a low glass transition temperature, T_g, organic matrix which has a degree of flexibility after forming. The amount of binder used depends upon the material which is being processed and its starting particle size. Usually 5–10% by weight of the total body composition is adequate to provide the 'green' strength necessary to produce continuous tapes. This is higher than the 2–5% normally used in

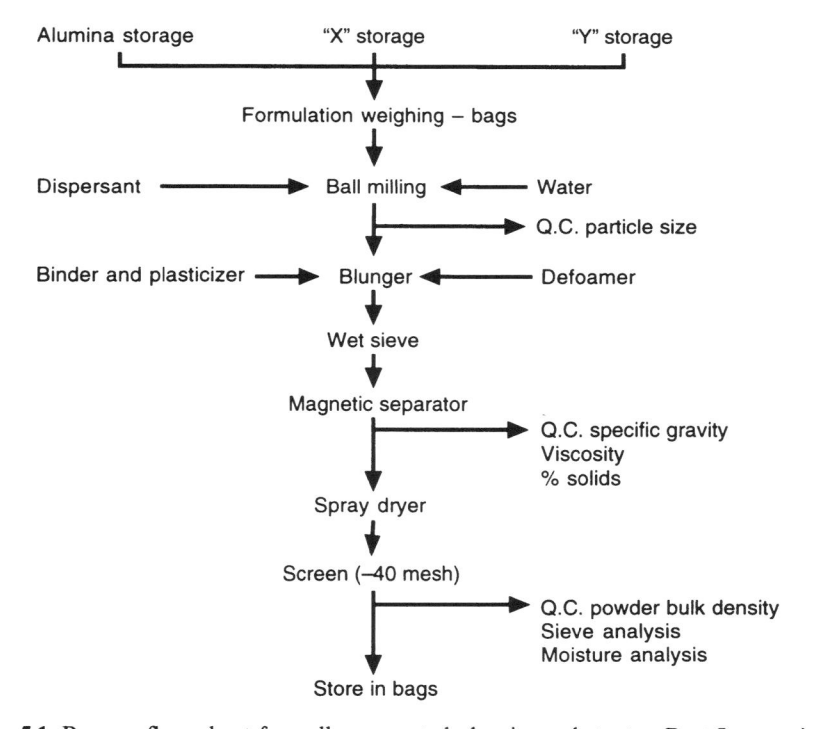

Fig. 5.1 Process flow chart for roll compacted alumina substrates. Part I: materials preparation.

a dry pressing operation. Another difference to be found when comparing the powders for roll compaction and those for dry pressing is in the moisture content. For roll compaction values as high as 7–13% are not uncommon, whereas for dry pressing the values are usually 0.1% or less. The last point to be made when comparing the two processes is in the size of the ultimate granules which are used as feed stock. For roll compaction a much wider range of spray-dried particles is utilized. Only the very coarse granules are removed (> 40 mesh or 425 μm). For dry pressing the standard range of sizes for the spray dried granules is −80 to +325 mesh or 45 to 180 μm. Figure 5.2 is a schematic representation of the actual fabrication steps in the roll compaction process. Feeding of the spray dried powders to the rolls is accomplished strictly by gravity. Force feeding by augers, etc. should not be used. The heart of the process is depicted graphically in Fig. 5.3.

The rolls turn in concert in opposing directions with identical tangential speeds. Very high pressures (in excess of 140 MPa) are generated along the line where the rolls come the closest to one another. Variable speed drive motors are utilized to control the rate of tape production. The speed of roll compaction is usually in the range 5 to 10 cm s^{-1}. Roll diameter is critical to the thickness of

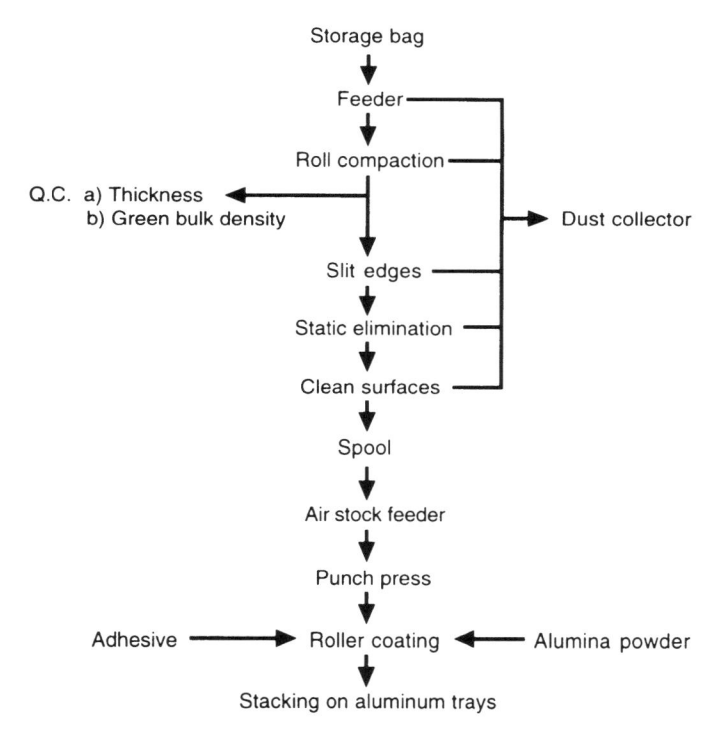

Fig. 5.2 Process flow chart for roll compacted alumina substrates Part 2: fabrication.

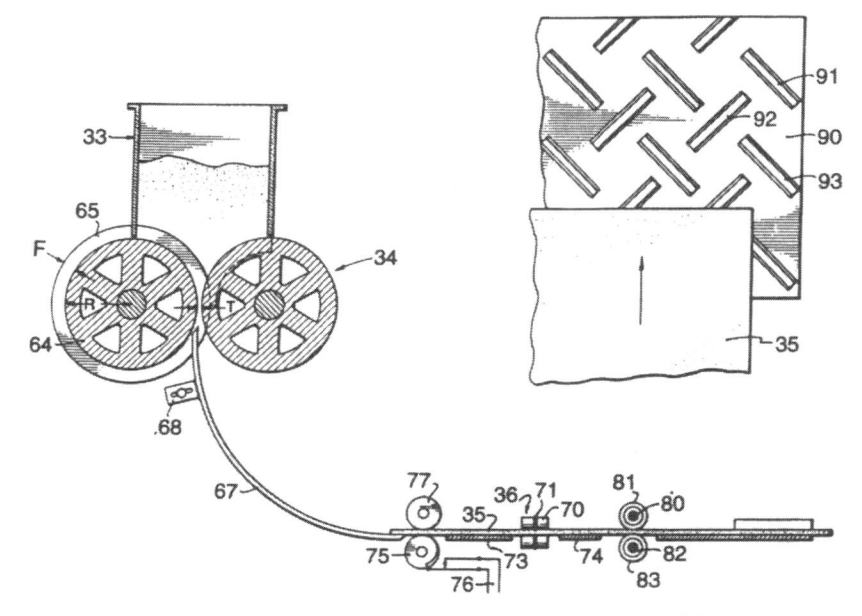

Fig. 5.3 Roll compaction schematic. (Source: Ragan [2].)

the final tape which can be produced. Generally, thicker tapes require the use of larger diameter rolls. Most production roll compactors have roll diameters in the range 150–250 mm. For a roll diameter of 200 mm, tape thicknesses in the range 0.100–2.29 mm have been produced. The very thin tapes (< 0.25 mm) are extremely difficult to manufacture by roll compaction. Since the compressive forces are so high, the unfired density is also very high and very uniform. As a result of this the shrinkage during sintering is relatively low and very uniform. Some of the advantages of roll compacted tape are evident on Fig. 5.2: the ability to spool the 'green' tape and the ability to punch to size and shape. The roll coating denoted on this diagram is to apply an adhesive (usually polyvinyl alcohol) and a fine alumina powder as a parting layer so that the substrates can be stacked during sintering. Figure 5.4 is a schematic of the sintering, cleaning and final inspection of the alumina substrates produced by roll compaction. These steps are very conventional and are used in many ceramic manufacturing lines. The major advantages of roll compaction for the production of ceramic substrates when compared with tape casting are as follows:

- water-based technology,
- easily incorporated into a factory with a spray drying facility,
- more akin to conventional ceramic processing,
- low, uniform shrinkage during firing, and
- excellent for lower quality surface finish substrates, e.g. 90–96% alumina.

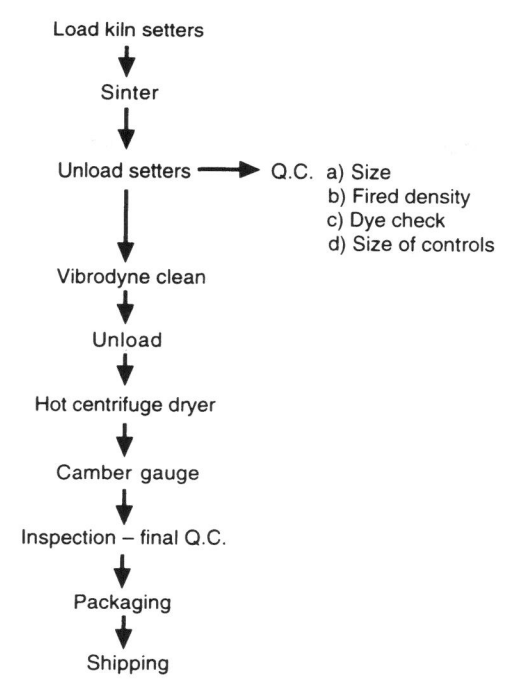

Fig. 5.4 Process flow chart for roll compacted alumina substrates. Part 3: sintering and inspection.

The disadvantages of roll compaction are:

- need a spray dryer if not in a conventional ceramic manufacturing factory,
- cannot produce very thin tape (< 0.25 mm),
- surface quality not adequate for thin film substrates, and
- tapes cannot be laminated for multilayers.

Tape casting, doctor blade casting and knife casting are different names for the same process. The 'doctor' is a scraping blade for forming a thin, basically two dimensional, structure such as the alumina substrate shown in Fig. 5.5. Two dimensional structures are defined as having a large X and Y area and a very thin, usually 0.635 mm or less, cross section or thickness. The range of thicknesses which have been tape cast by the author actually exceed this value and extend from 0.025 mm to over 3 mm. Tape cast products range from small 1.5×1.5 mm square thermistors to sheets as large as 1.2×2.4 m square which are used in fuel cells. In the ceramics industry, tape casting is a process which is most akin to slip casting. That is, it is a forming process from a fluid suspension of ceramic particles. There are subtle differences, however. The suspension is usually non-aqueous because tape casting involves an evaporative drying process rather than an absorption process into a porous plaster of Paris mould.

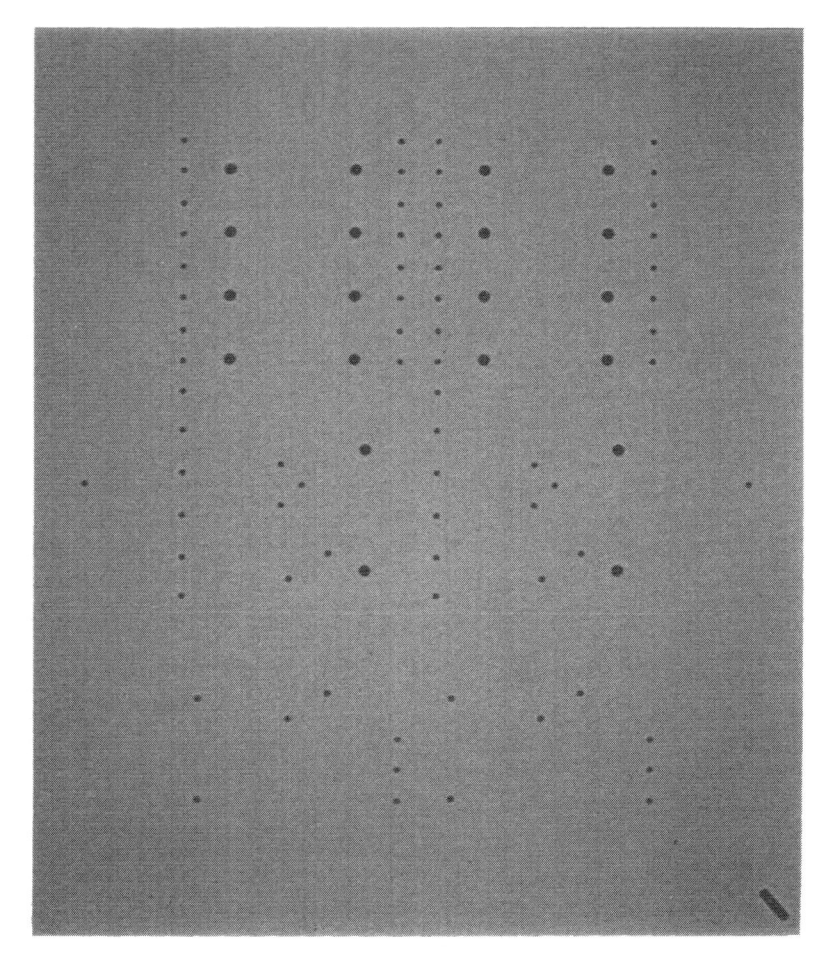

Fig. 5.5 Tape cast aluminum oxide substrate with punched holes.

Tape casting was originally developed by Glenn N. Howatt as part of an effort to produce thin piezoelectric materials during the Second World War in the mid 1940s. His first publication and patent filing took place in 1947–48 [3, 4]. Howatt's patent taught the doctor blading of a slurry onto plaster slabs where the solvent, which could be water, was partially removed by absorption. The slabs were on a moving conveyor which was synchronized with the flow of the slip. An oven dried the top of the cast pieces and 75% of the solvent was removed before the pieces could be stripped from the carrier and transferred to a drying and sintering furnace. In the mid-1950s John L. Park, Jr. of the American Lava Corporation filed a patent which refined the Howatt patent and taught the casting of a non-aqueous based slurry onto a moving polymer carrier such as polyester [5]. Refinements of the basic process described by Howatt and Park have been numerous, however the initial processing steps outlined by these authors still apply today.

The remainder of this paper will describe the process, materials, and applications of tape casting as it is used today in many ceramic, metallic and combined products.

5.2 THE DOCTOR BLADE TAPE PROCESS

Modern tape casting technology is based upon the Park patent. That is, the casting usually is done on the surface of a moving polymer carrier or substrate. Figure 5.6 illustrates the basic principles of tape casting. In this case the carrier is a cellulose acetate film onto which a layer of slip is 'doctored' by a blade as the carrier film moves under it. The simplicity of the process is what makes it so intriguing to the ceramic engineer who is interested in fabricating thin sheets of ceramic products. The design of doctor blades, casting machines and the formulation of casting systems are not as simple as the basic process looks at first glance. These are the topics which will be reviewed in this chapter and which are based upon over twenty years of personal experience by the author.

5.2.1 Materials technology and selection

Other than the ceramic powder, the most important choice which must be made is the selection of a binder/plasticizer/deflocculant system. To produce a flexible tape cast product the binder/plasticizer must be a long chain polymer

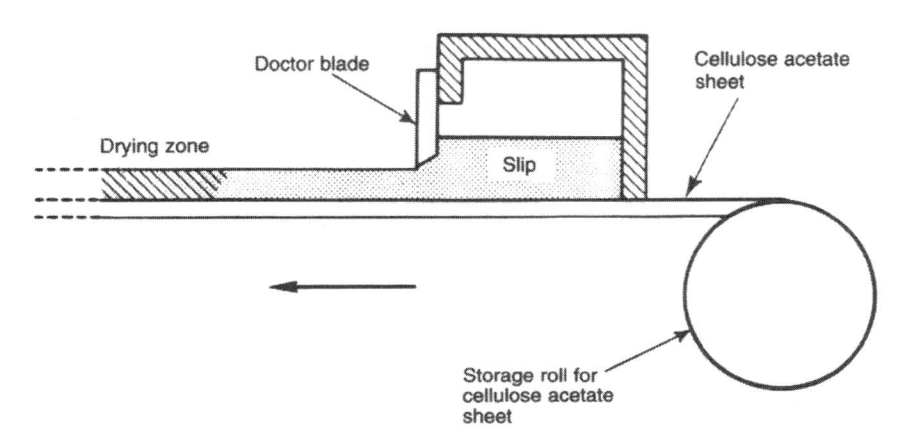

Fig. 5.6 Doctor blade casting head schematic. (From: Mistler *et al.* in *Ceramic Fabrication Processing Before Firing*, (eds Onoda and Hench) 1978. Reprinted by permission of John Wiley & Sons.)

and it must be a 'film-former' when dried from a solvent system. The selection of a binder/plasticizer is further complicated by the conditions set for removal before sintering. Most oxide ceramic materials can be sintered in an oxidizing atmosphere. This makes the binder removal step easy since most organics burn in the presence of oxygen and heat. On the other hand, if the matrix powder is a reactive metal, carbide, or nitride the binder removal must be accomplished in a neutral or reducing atmosphere. This puts more of a restriction on the binder selection and would dictate the use of an organic system which 'un-zips' or evaporates upon heating. Commonly used binders for oxidizing firing include poly(vinyl acetate), poly(vinyl chloride), poly-styrene, poly(vinyl butyral) and various co-polymers. The acrylics are used under non-oxidizing conditions. In a recent publication [6] a complete list of organic additives that have been used in tape casting is tabulated along with the reference work for each system. In addition to the restrictions already mentioned, other important factors which must be taken into consideration are cost of the binder and solubility in an inexpensive, volatile and relatively non-flammable solvent. The use of a non-flammable solvent is almost imposs-ible to accomplish unless one chooses to use the chlorinated hydrocarbons and then health restrictions come into play.

There are other considerations which must be made when selecting a binder/solvent system. For example: the final tape thickness is a factor in the selection of a solvent. Usually for thin tapes (those <0.254 mm) a highly volatile solvent such as acetone is desirable whereas for thick tapes (>0.254 mm) a lower volatility solvent such as toluene or xylene can be used. Another consideration is the casting surface; on what will the slurry be cast? Is there a reaction between the solvent or binder and the casting surface? For example; poly(vinyl butyral) adheres strongly to glass, therefore it cannot be used as a binder if glass is the casting surface of choice. Solvent selection can be limited by health and environmental considerations. The use of chlorinated hydrocarbons such as trichloroethylene has largely been eliminated in most tape formulations as a result of this. Safe solvents such as the alcohols and combinations of alcohols with toluene, xylene and methyl ethyl ketone are currently in vogue. The ultimate goal of many researchers in recent times has been to use aqueous-based tape casting systems. This would alleviate the health and environmental problems associated with non-aqueous solvents but a host of other problems crop up with water-based systems. Some of these include:

- curdling of the slip due to pH instability,
- foaming in the mill,
- premature polymerization,
- foaming during de-airing,
- cracking during curing,
- tape brittleness,
- excessive curl during cure,

- poor tape surface quality,
- 'fish eyes' or de-wetting marks,
- stress marks on the tape during release, and
- cracking during firing.

Many aqueous systems have been developed and tabulated [6]. Several of these have been tried by the author and have not worked as stated.

Plasticizers are added to tape systems to extend the flexibility of the dried tape. This is accomplished by modification of the glass transition temperature (T_g) of the polymer binder. The list of plasticizers is extremely long and most researchers turn to recommendations of the binder manufacturers in initial formulations. In some of the authors' experience it has been found that a dual plasticizer is beneficial, i.e. a T_g modifier and a plasticizer for the tape itself, not a polymer modifier. The common plasticizers for non-aqueous and aqueous systems are also tabulated [6]. One important consideration to be made in the selection of all of the organics in a tape system is the mutual solubility and/or compatibility of the ingredients. All of the organics should be soluble in the solvent system selected. This will prevent phase separation, etc. This has been emphasized by several authors [5, 7].

Deflocculation/dispersion as in all ceramic processing technology is critical to tape casting. Slip solids loading, slurry viscosity and rheology, tape packed density and tape homogeneity are all influenced by the state of dispersion of the ceramic particles in the slurry. Many of the advances made in tape casting have had their roots in the paint literature and patents. In organic systems the predominant mechanism which provides the dispersion is known as steric hindrance where the long chain fatty acid type of organics, e.g. glycerol trioleate or fish oil, attach themselves to the ceramic particles with long tails pointing outward and prevent re-agglomeration by mechanical separation of the particles. In aqueous systems the standard electrostatic mechanisms which have been described by many authors also are active and are pH controlled. Many aqueous systems have combined steric hindrance and electrostatic dispersion mechanisms. Several excellent papers have been published in the past five years which deal with dispersions in tape systems [8, 9]. The common dispersion/deflocculant agents for aqueous and non-aqueous systems are listed [6]. A thorough discussion of deflocculation has previously been published by the author [10].

A procedure for the determination of the optimum percentage of deflocculant which should be added to a non-aqueous solvent–ceramic system was devised and published by Tormey [11] as part of her thesis work. Basically the procedure involves the following steps (when used with fish oil or glycerol trioleate):

1. prepare a colorimetric calibration curve of absorbance against concentration of dispersant,
2. mill dispersant, solvent and ceramic powder for 20–24 h,
3. centrifuge,

4. analyse supernatant solution colorimetrically,
5. determine concentration of dispersant in solution from calibration curve,
6. calculate the concentration of dispersant attached to the ceramic particles (by difference),
7. plot absorption versus wt% dispersant in solvent, and
8. the curve developed will level-out to an equilibrium concentration.

A concentration just beyond the break in the curve on the flat portion is the proper amount of dispersant to add for the system tested.

A similar type of test, which is easier to perform has also been outlined by Tormey [11]. This consists of the following steps:

1. mill dispersant, solvent and ceramic powder for 20–24 h,
2. pour into a 10 ml graduated cylinder with a stopper,
3. let settle until settling is complete,
4. record the height of settled bed and condition of supernatant liquid,
5. repeat at several concentrations of dispersant, and
6. plot a curve of settled bed height against dispersant concentration.

The usual shape of this curve is for the bed height to decrease as a function of dispersant concentration and then to reach an equilibrium level at the optimum concentration. This would be the level of dispersant to use for this solvent/powder combination. The author has used this technique for several years in formulation experiments. Typical formulations for non-aqueous and aqueous tape systems are listed in Tables 5.1, 5.2 and 5.3.

5.2.2 Tape casting process and equipment

(a) Slip preparation

Standard techniques used for decades in the ceramics industry are utilized to prepare the slips or slurries for the actual fabrication step-doctor blade casting.

Table 5.1 Typical non-aqueous tape casting formulation for use in an oxidizing atmosphere sintering process

Formulation component	Weight (%)
Aluminum oxide (ceramic powder)	67.4
Fish oil[a] (dispersant)	1.2
Anhydrous ethyl alcohol (solvent)	9.2
Xylene (solvent)	14.2
Mixed normal alkyl phthalate[b] (plasticizer)	2.4
Polyakylene glycol[c] (plasticizer)	2.9
Polyvinyl butyral[d] (binder)	2.7

[a]Z-3 Grade, Reichold Chemicals, Inc., White Plains, NY, USA.
[b]PX-316 Grade, Aristech Chemical Corp., Pittsburg, PA, USA.
[c]UCON 50 HB 2000, Union Carbide Corp., Danbury, CT, USA.
[d]B-98, Monsanto Corp., St. Louis, MO, USA.

Table 5.2 Typical non-aqueous tape casting formulation for use in non-oxidizing atmosphere sintering processes [24]

Formulation component	Weight (%)
Barium titanate (ceramic powder)	69.9
Fish oil[a] (dispersant)	0.7
Methyl ethyl ketone (solvent)	7.0
Ethyl alcohol (solvent)	7.0
Butyl benzyl phthalate[b] (plasticizer)	2.8
Polyethylene glycol (plasticizer)	2.8
Cyclohexanone (homogenizer)	0.5
Acrylic in methyl ethyl ketone[c] (30% solution) (binder)	9.3

[a]Z-3 Grade, Reichold Chemicals, Inc., White Plains, NY, USA.
[b]Santicizer 160, Monsanto Corp., St. Louis, MO, USA.
[c]Acryloid B-7, Rohm and Haas Corp., Philadelphia, PA, USA.

Table 5.3 Typical aqueous-based tape casting formulation [25]

Formulation component	Weight (%)
Mullite powder (ceramic powder)	56.4
Darvan C[a] (dispersant)	0.7
Water (solvent)	25.8
Polypropylene glycol[b] (plasticizer)	2.9
Acrylic emulsion[c] (binder)	14.3

[a]R.T. Vanderbilt Co., Norwalk, CT, USA.
[b]PPG, $MW = 1200$, Fluka Chemical Corp., Ronkonkoma, NY, USA.
[c]Rhoplex B-60A, Rohm and Haas Corp., Philadelphia, PA, USA.

The ingredients are weighed and batched together in a mixing and milling container such as a standard ceramic ball mill. Usually the slip preparation process is divided into two stages as follows:

(i) Dispersion milling

During this first stage the ingredients which are added to the ball mill are: the solvent(s), the powder(s), the dispersant/deflocculant, and the grinding media which usually fill half of the empty mill before adding the other materials to be milled. The order of mixing the ingredients in a given formulation is critical to the final success of the slurry produced. Usually the dispersant/deflocculant is dissolved in the solvent system and added to the mill first. Then the powders are added and the mill is sealed and rolled or vibrated for a period of time from 4 to 24 h. The dispersion milling step serves several important functions.

Among the most important are: the breaking down of agglomerates, the breaking down of large particles into smaller ones, and last but not least the equal distribution of the dispersant/deflocculant on the surfaces of the individual particles created. This yields a fluid, low viscosity slip which is well dispersed in the solvent system.

(ii) Slurry mixing

Once the particles are suspended in the solvent then, and only then, should the plasticizer(s) and binder be added to the mill. Most manufacturers add the plasticizer(s) first followed by the binder. The reason for this is the simple fact that in many cases the binder is more soluble in the plasticizer than in the solvent(s) and it goes into the solution more readily when added in this order. Some binders such as poly(vinyl butyral) act as secondary dispersants with some of the inorganic powder systems and immediately begin to compete with the dispersant for the powder surfaces. This was the primary reason for separating the dispersion stage and the mixing stage of milling. The slurry begins to become more viscous during this mixing stage and the binder (if in a powder form) dissolves into the mixture. Some manufacturers prefer to predissolve any powdered binders and do not depend upon the mill to provide the solution. Standard practice is to mill the binder/plasticizer(s) into the slurry for 12–24 h.

The next step in slip preparation for tape casting is to remove any entrained air bubbles from the slurry. This can be done by slow rotation for 24 h or more on a set of rollers (usually at less than 10 rpm). By far the most common technique is to use stirring accompanied by the application of a vacuum in a closed container such as a tank or desiccator. The vacuum applied is usually very little, 635–700 mm Hg. Too high a vacuum will tend to remove a large volume of solvent along with the bubbles. Depending upon the volume of slip to be de-aired the time required will range from 8 min to 1 h. As a rule-of-thumb 4 litres of slip will usually be de-aired in about 5–8 min under these conditions. At this point in the manufacturing scheme a slip characterization sequence is usually inserted for in-process quality control. The characteristics which are monitored are:

- viscosity, standard Brookfield viscometer can be used with an RV4 spindle at 20 rpm. Values in the range 500–6000 mPa s (cP) are common;
- specific gravity, weight of a graduated cylinder of slip, usually 100 ml;
- particle size, diluted slip can be used with any analyser that can tolerate solvents.

After de-airing, the slip can be used directly for tape casting, however when defect-free tapes are required or when very thin tapes are being produced a filtration step can be introduced. A typical filter has been described in the literature [12]. Basically, the slip is pumped or pressure-fed up through a series of nylon (or other non-reacting material) screens. The mesh size of the screen openings is selected based upon the particle size of the milled powders and the

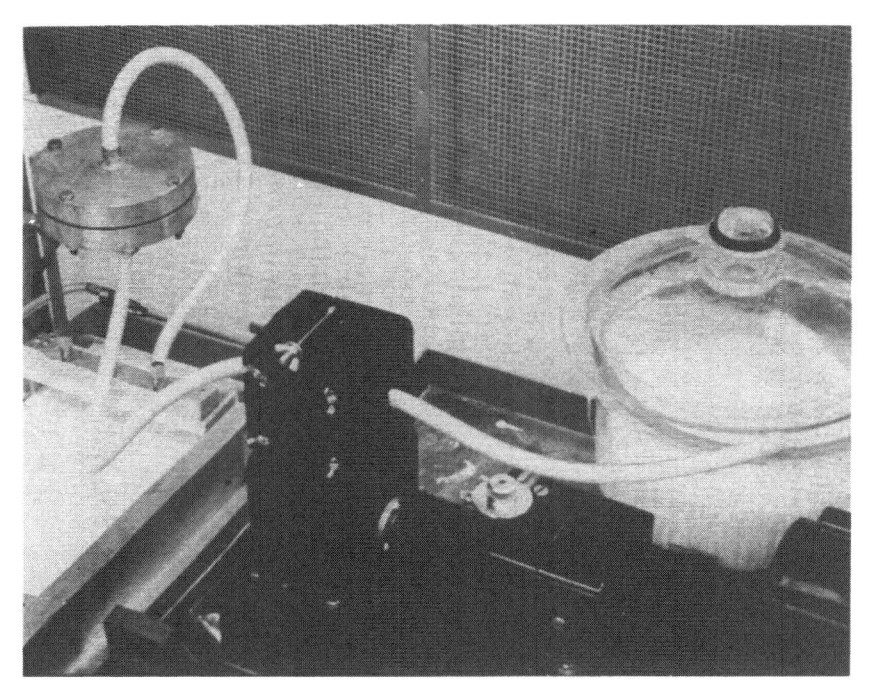

Fig. 5.7 Experimental vacuum de-airing and filtration system for tape casting. Reproduced from [6].

viscosity of the slip. Mesh openings as small as $10 \, \mu m$ have been used successfully with aluminum oxide slurries at $2500-3000 \, mPa \, s$ viscosities. The filter removes defect promoters such as pieces of undissolved binder, pieces of the mill or grinding media, powder agglomerates which have not broken down, and any remaining air bubbles which were not removed during de-airing. It is important to remember to pump the slip **up** through the filter, thereby pushing any air ahead of it before going through the mesh itself. Figure. 5.7 is a photograph of a vacuum de-airing desiccator, pump, and filter system at the casting head section of a tape casting machine.

(b) Tape fabrication

After the slip is filtered it is delivered directly to the working section of the tape casting machine, the doctor blade. A doctor blade is simply a gating device to control the wet thickness of the slurry being applied to the surface of a substrate carrier. Two modes of operation are in use today: a moving blade–stationary carrier system and a stationary blade–moving carrier system. For mass production and continuous operation the stationary blade system is preferred. The blade height is set a specified distance above the carrier by means of dial micrometers. Most doctor blade assemblies consist of one or more adjustable stainless steel blades in an aluminum alignment frame which also serves as a slip reservoir behind the blade. If the carrier is plastic the doctor blade assembly can ride directly on its surface. If the carrier is glass or steel it is advisable to provide PTFE tape covered runners on the supports for the doctor blade. This will help to eliminate friction and scratching of the carrier during use.

The design of doctor blades and doctor blade assemblies is more of an art than a science. The shape and size of the doctor blade itself can have a profound influence on the tape which is formed and therefore the final product. Shapes which have been described in the literature range from a simple square bottom to a bevelled knife edge. Other workers prefer a parabolic contoured shape. The author has used square-bottomed blades for over 20 years with excellent success. The bottom width of the blade has also received considerable attention. Widths ranging from 6–7 mm to 25 mm have been used. The author prefers the narrower blade simply because of the added weight which results from the wider steel blade. A recent publication [13] has attempted to put some science into doctor blade design.

For precision tape casting where tolerances of ± 0.0025 mm must be maintained over a 200 mm width and along a continuous length of tape most researchers and manufacturers use a double doctor blade. The double doctor blade provides better control of the hydraulic 'head' behind the lead or casting blade. A schematic of this principle is shown in Fig. 5.8 which is taken from a publication by Runk and Andrejco [14]. In their paper they explained that the best results were obtained by setting both blades at exactly the same height or opening above the carrier.

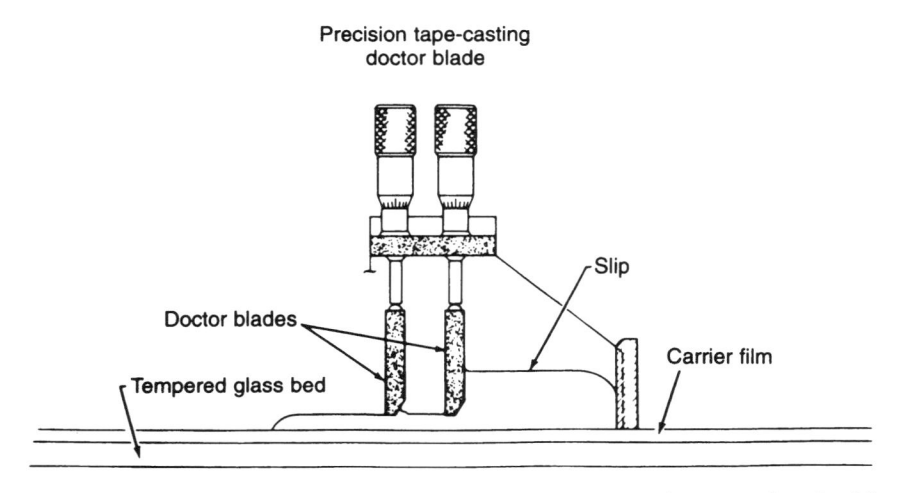

Fig. 5.8 Precision tape casting double doctor blade schematic. Reproduced with permission from [10].

The role of the doctor blade is to meter the slurry onto the carrier to the exact thickness desired. What is that thickness? It is complicated by the fact that one is metering a slurry which dries to an unfired tape which sinters to a solid ceramic part. Each of these steps involves shrinkage. For example: an alumina slurry as reviewed previously in Table 5.1 when cast with a blade gap setting of 1.57–1.60 mm will dry to an unfired tape of 0.76–0.81 mm. Upon sintering this tape will shrink to a final thickness of 0.635–0.69 mm.

Thickness control is further complicated by the fact that it is a function of slip viscosity, casting speed, and reservoir depth in addition to the simple act of setting the blade gap. Once a set of parameters are developed for a given system and the variables are held constant thickness control can be achieved. In many cases trial runs are the only way to determine the proper settings for the variables mentioned above. One method for controlling and monitoring thickness of the cast slurry is to control the depth of the slip behind the doctor blade. This is usually accomplished by using a sensor of some sort (ultrasonic, photonic, air or capacitance) to detect the slip level and to activate an electrical valve to control flow through the feed tube(s). A γ-ray back-scatter apparatus can be used to monitor the 'wet' tape thickness 'on-the-fly'. This instrument can also be used to activate a control system on the reservoir depth. Changing the doctor blade height while the casting process is underway is not recommended as a control procedure.

In a stationary blade system the slip is pumped into the doctor blade reservoir and the carrier is set in motion. The casting speed is usually pre-set either for a continuous cast or a batch cast. The exact speed of casting depends upon the length of the drying chamber on the machine and it also depends upon the thickness of tape being cast and the solvent system being used. Typical speeds for casting range from 15 cm min^{-1} for a continuous cast to 50 cm min^{-1} for a batch cast. Some material systems do not fall into that range, e.g. the author has observed thin capacitor tapes being cast at rates which can only be measured in metres per minute. This is not standard practice in most ceramic systems.

The major components of a tape casting machine can be seen in Fig. 5.9, which is a pilot level to small scale production machine. The machine consists of a solid, level casting surface, a drying chamber with a built in adjustable speed exhaust fan to draw air counter current to the movement of the carrier and tape, an adjustable speed control for the carrier and an air heater to control the temperature of evaporation and drying. Some machines are also equipped with platten heaters below the casting bed to provide heat from the bottom of the tape.

Air flow through the machine is essential to evaporate and carry away the solvents during drying. Heated air facilitates this and heated plattens permit rapid drying even when the drying process becomes diffusion controlled. Diffusion becomes rate controlling when the liquid slurry gels and begins to solidify. Laminar air flow is usually recommended in most tape casting processes. For most alcohol based systems this would be less than 3 m^3 h^{-1}.

Fig. 5.9 Tape casting machine. Photograph courtesy of Unique/Pereny Furnaces, Kilns, and Dryers, Ringoes, NJ, USA.

The kinetics of the drying process have been described previously [10]. A typical drying curve which exhibits the two stages involved, i.e. evaporation followed by diffusion control, is schematically shown in Fig. 5.10.

There are numerous casting surfaces which are used in continuous tape production machines. They can be categorized as: organic, coated organic, metal and coated paper. By far the most common casting surfaces are polyester and coated polyester. Several machines have been built with a continuous stainless steel belt, however the organic polyester carrier is used in more production situations. Which surface to use is determined by the interaction of the binder/solvent system with the carrier material and the resultant release characteristics. If formulated correctly many slurry systems can be cast directly onto stainless steel belts with a polished surface. The author has had success with numerous binder/solvent systems and materials on silicone-coated Mylar [15]. In most cases the carrier selection is made by trial and error during the development stage and the production process is built around the material chosen at that time.

Casting machines are very simple in design since they are really just elongated forced air drying ovens. The addition of multichambered drying with separate temperature controls (and at times separate air flow controls)

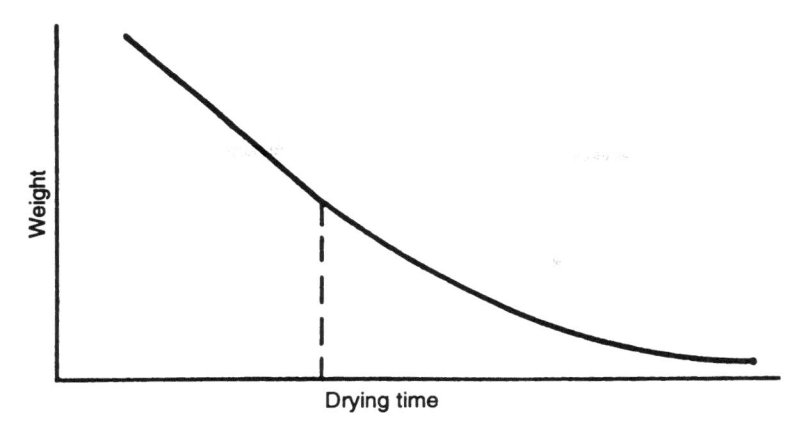

Fig. 5.10 Schematic of weight against time in a two-stage drying process.

can produce a wide variety of temperature–time profiles in the machine. When designing a casting machine one must keep in mind that its primary purpose is to remove the solvent(s) as fast as possible without exceeding the boiling point of the organics and without forming a skin on the surface at too early a stage of drying. The creation of a 'case hardening' effect can occur if the solvent gradient and/or temperature gradient is not adjusted properly. The ideal condition is to have a high solvent vapour concentration at the entrance end of the machine near the casting head and a continuously decreasing concentration as you approach the exit end. This is accomplished by placing the exhaust fan or duct just downstream from the casting blade assembly. The ideal temperature gradient is to have room temperature at the casting head and a continuously increasing temperature as you approach the exit end of the machine. This is accomplished by placing the air heating assembly near the exit end of the machine. Most machines have a 'dead space' designed into them beyond the heater so that the tape can cool down before removal from the carrier. The peak temperature should be reached at the point where the majority of the solvent has been removed from the then almost dry tape.

The next step in the tape casting process is the removal and handling of the dried tape. In a batch operation the tape is cast and moved into the drying section of the machine. Once the tape is dry it can be run out of the machine, stripped from the carrier using a metal blade, and cut into lengths for storage or punching. In a continuous operation the tape is cast and dried while the carrier continues moving through and out of the exit end of the machine. In this case the most commonly used technique is to strip the tape and roll it on to a take-up spool. One can interweave a separation layer of material such as paper to prevent blocking (sticking to itself) of the tape. The carrier upon which the tape was cast is then wound on to a separate spool and is either discarded or reused.

(c)　Tape optimization and characterization

The optimization of the organic content of the tape cast product is essential since the shrinkage, densification, and warping tendency all depend upon this important parameter.

Thermal gravimetric analysis (TGA) of microtomed sections of the tape taken from the top and bottom can be compared to determine the uniformity of the binder/plasticizer distribution in the tape. If too much binder/plasticizer is present or if there is segregation the TGA will detect it. Figure 5.11 is an actual set of TGA scans from the top and bottom of an alumina tape with a poly(vinyl butyral) binder. A degree of segregation is exhibited with a greater percentage of burn-off from the top layer. This indicates that some settling of the inorganic phase has taken place with a higher concentration of the organic phase at the top surface. The ideal tape would exhibit little or no difference from top to bottom. A condition such as shown on this figure would, in all probability, lead to a warped part upon sintering. The warpage is caused by the gradient of inorganic particle packed density from top to bottom of the tape. Other than physical measurements such as thickness, the most common characterization tool for 'green' or unfired tapes is the 'green' bulk density. The most common technique used is to measure physically the volume of a piece of

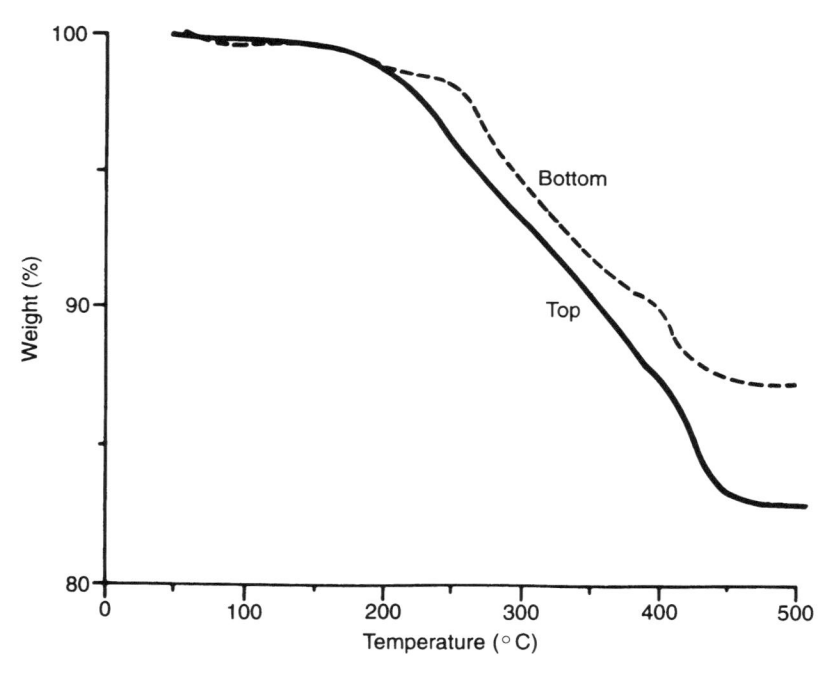

Fig. 5.11 Thermogravimetric analysis of microtomed sections taken from top and bottom surfaces of an alumina tape cast ceramic.

tape of known dimensions by measuring the thickness, using flat-faced micrometers at several points, and the area of the piece. The part is then weighed and the 'green' bulk density is determined by dividing the weight by the volume. Tape cast ceramics exhibit very high bulk densities, usually 55–65% of the theoretical density of the organic matrix phase. For aluminum oxide tapes values as high as $2.55\,\mathrm{g\,cm^{-3}}$ are common. The reason for such high values in a non-pressurized process such as tape casting is the fact that well dispersed particles tend to stay in suspension even during drying and therefore pack together better as the solvent is removed and the particles finally settle into a dense 'green' tape. Sinterability is dependent upon how well the particles are packed together. Poor deflocculation leads to poor packing which ultimately leads to a low sintered density.

5.3 SUBSEQUENT PROCESSING OF UNFIRED TAPES

5.3.1 Shaping

After the tapes are formed the next step in processing is to cut the parts to their final stages by punching. In most full-scale manufacturing operations this is accomplished by using punch and die sets in either a hydraulic or mechanical press. There are actually two separate and distinct functions which can be performed during the punching operation.

(a) Blanking

Blanking is the step which forms the outside shape of the part. For ceramic substrates this is usually a simple square or rectangular geometry.

(b) Hole or via generation

This is the means to locate small circular or square holes in the 'green' tape. For multilayered ceramics the via holes are used to interconnect the various layers with plugs of metal conductors.

When punching unfired tapes it is essential to make sure that the punching tools are kept clean of any debris which forms from the punching itself. Usually an air jet or vacuum cleaning system is built into the tools. In critical applications where very clean, defect-free substrate surfaces are required, e.g. thin film substrates, punch faces are designed with recessed surfaces so that punching contact is only made at the edges of the substrate blank.

Hundreds of via holes can be punched simultaneously with the blanking operation. For multilayered ceramic tapes registration holes are also punched at the same time. Registration holes are usually 5–7 mm in diameter and are used to align the different layers during the subsequent processing steps.

Via holes > 0.5 mm in diameter are usually very easy to form since materials such as tungsten carbide can be used as punching tools. For holes as small as

0.25 mm or less tungsten carbide may be too brittle and other materials such as borided tool steel can be used. There are companies which will boride punches which have been machined to size. The boriding process does not change the final size of the punch since it is a diffusion process. By using borided steel punches the surface is as hard as carbide but has the flexibility of steel.

One must remember that precision location of holes is difficult in a tape cast product since the shrinkage during sintering is large and different in the as-cast and cross cast direction. An excellent paper is in the published literature which deals with this problem in detail [16].

The design of punches and dies for tape cast ceramics has been covered in several publications by the author and will not be repeated here [10, 17]. One important criterion is that the punch is always made of a harder material than the die itself, e.g. a carbide punch would be used in a tool steel die. Punches are designed to last for 10–100 thousand strokes.

5.3.2   Multilayer processing

Multilayered ceramic packages and multilayered capacitors are based upon tape casting technology. The two predominant processing techniques which are used for the production of multilayers are lamination and built-up screen printed layers. These will be described in detail in the sections which follow.

(a)   Laminated–cosintered multilayers

The major process which is used in the multilayered ceramic package industry today is based upon work which was performed at RCA (Radio Corporation of America) in the late 1950s and early 1960s. Stetson and Schwartz outlined the basic principles for this technology in a paper presented in 1961 [18]. In that paper they pointed out the primary advantage of coincident sintering of the ceramic and metal phases in a multilayered structure: one sintering step produces a monolithic ceramic structure with buried metallic interconnections. The technology has grown to include multilayer capacitors, ferrite memories, semiconductor packages and multilayer ceramic packages. The multilayered ceramic technology introduced and manufactured by IBM is the state-of-the art today and is used in their most sophisticated computer systems. Some of their packages have over 30 layers laminated and sintered into one monolithic package. The basic steps involved in the laminated ceramic process are:

1. tape cast sheets,
2. punch holes,
3. print patterns,
4. laminate sheets,
5. punch final shape, and
6. sinter.

The first two steps have already been discussed above, the remaining steps will be discussed below.

(i) Pattern printing and via metallization

In this section the materials systems, formulations, and processing for multilayer thick film inks and the screen printing and via filling processes will be discussed.

Materials systems Most metallization systems, or 'inks' as they are commonly called, include the following ingredients:

- metallic phase – electrical conductor,
- flux phase – to reduce the sintering temperature,
- dielectric and/or glass phase – bonding agent to ceramic,
- organic phase – to provide unfired strength and screenability, and
- solvent phase – to provide viscosity control.

Not all of the ingredients listed are necessary in all formulations but as a general rule they are usually present. There are not many references in the published literature which review metallization systems. Most companies which use cosintering technology prepare their own tape systems and matching metallization systems themlseves and the formulations are regarded as proprietary.

The selection of a metallic phase depends upon the sintering conditions which will be encountered during processing. Commonly used metals include: silver, gold, copper, palladium, platinum, rhodium, iridium, molybdenum and tungsten. The temperature and atmosphere of cosintering will limit the selection of the metal, e.g. for aluminum oxide and beryllium oxide ceramics molybdenum and tungsten based inks must be used because of the high temperatures required. On the other hand recent developments with ceramic multilayer systems which sinter at or below 1000 °C can use silver, gold or copper based inks.

Other considerations to be made in the selection of a metallic phase include thermal expansion match with the ceramic, cost, and conductivity of the metal.

The flux phase is usually included to optimize sintering of the primary metal by forming a liquid phase at the eutectic temperature. Common fluxes for molybdenum include manganese and titanium hydride. Nickel is a fluxing agent for tungsten. Most of the refractory metals would sinter at a temperature higher than the sintering temperature of the ceramic phase if a flux was not included.

Probably the most important ingredient in a metallization formulation other than the metal itself is the dielectric phase which is added to provide better adhesion to the ceramic substrate and the via walls. A secondary purpose for the addition of a dielectric phase is to provide a better match of thermal expansion between the metallization and the vias and/or the substrate surface. A dielectric phase also provides some match of shrinkage during cosintering of the metallization and the substrate. An IBM patent issued in

1982 uses this technique to eliminate via to via cracking in a multilayer package [19]. Since most manufacturers prepare their own metallization inks it is not widely known that the inks incorporate as much as 15 wt% of the substrate material to provide bonding and a better match of thermal expansion. Common materials used as dielectrics are glass frits, raw ceramic oxides, or combinations of the ingredients in the substrate itself.

There are numerous organic vehicles which are used to provide 'green' or unfired strength in the metallization. By diluting the organic phase with solvent the viscosity of the ink can be tailored to the coating process being used. For example a screen printing ink would have a viscosity of 100 000–200 000 mPa s (cP) whereas an ink for application by spraying or painting would have a much lower viscosity of the order of 1000 mPa s (cP). Typical binders used in metallization inks include ethyl cellulose, nitrocellulose and methyl methacrylate. Solvents include terpineol, butyl carbitol acetate, and butyl cellulose acetate. This is just a short list of organic binders and solvents which can and have been used in metallization inks.

Materials processing   The first step in the preparation of metallization inks or paints is mixing and milling of the metal powders, fluxes, and dielectrics. The materials are weighed and added in the proper proportions to a jar or ball mill. The most commonly used mills are steel with steel grinding media. If iron contamination is a problem aluminum oxide mills with aluminum oxide balls or flint pebbles can be used. Acetone is widely used as a grinding medium since it can be dried rapidly after the milling is complete. Grinding times range from 24 to 48 h or until the proper size reduction is reached. After the milling and drying are complete the powder can be checked for particle size distribution and surface area.

When preparing a screen printing ink or paste the second mixing procedure is usually done in two stages. First, the powders prepared by milling are mixed with an organic binder and solvents in a low shear type mixer. Then the resultant paste is passed through a three roll ink mill at least three times. The ink mill smears out the organics and the solid particles to form a smooth viscous paste. At this point the screen printing ink can be collected and stored in a closed container.

Ink characterization   At this point another series of characterization checks are made on the finished metallization ink.

The first of these is a check of the fineness of grind using a Hegman Gauge. The Hegman Gauge is a simple device to give an indication of how well the individual agglomerates have been broken down. A doctor blade type blade is drawn across a wedge-shaped plate and a number is assigned to the fineness of grind depending upon where the first agglomerates of material show streaking. This is a procedure which is commonly used in the paint industry.

Viscosity is checked using a standard viscometer. This can be either a cone and plate type or a common spindle type apparatus.

A standard proof test to evaluate metallization inks is to screen print the ink onto the end of a cylindrical piece of a standard ceramic composition which has previously been sintered. A standard lot of ink is processed along with the new lot of ink which is being evaluated. The metallization is sintered under identical conditions of temperature, time and atmosphere. The fired metallization is then electroplated and two pieces are brazed together end to end. The pieces are then tested in a standard three line loading jig by applying a force parallel to the braze line. The breaking force and the mode of breakage are recorded and compared with the standard ink joint values. Any technique such as this which evaluates the bond strength of the ink can be used as a proof test. ASTM test F19–64 is an excellent reference for such an evaluation procedure.

Metallization process   The usual steps involved in the metallization of unfired tape cast ceramics are as follows:

1. coat or fill vias,
2. print conductor patterns, and
3. dry.

The first process is to metallize the vias or through-hole interconnections. There are two procedures to accomplish this. The first of these is a process for coating the walls of the via only. For some applications, such as non-critical consumer electronics, this is adequate. The 'green' sheets of tape are stacked on top of one another so that the hole patterns are in exact alignment. This is one case for the use of the alignment holes which were punched around the periphery of the sheets. All of the pieces should be identical, i.e. one of the layers from a stack of multilayers. As many as 15 to 30 pieces can be stacked and metallized at the same time. A drop of ink is placed on a via hole on the top piece and a vacuum is drawn from the bottom through a vacuum plate. The ink is pulled down through the stack leaving a thin coating on the inside walls of the vias. A syringe or eyedropper can be used to deposit the droplet on the top piece.

For other applications where solid vias are required the procedure is quite different. Instead of using a thin fluid ink a viscous paste is utilized. As a matter of fact the via filling can be accomplished at the same time as the conductor patterns are deposited by screen printing. In this procedure the squeegee is drawn across the screen or stencil and the mass of ink is forced into the via hole. With large via holes more than one pass of the squeegee may be required. A vacuum is usually applied through a paper filter at the base of the hole to assist the filling of the via. Recently an injection system for filling via holes has been described [20].

After the via holes have been filled, or when they are being filled if done by a squeegee technique, the conductor line patterns are screen printed onto the green sheets. It is beyond the scope of this paper to go into the details of screen printing technology. For information see [21]. This paper reviews screen design and performance criteria. Some of the specific suggestions for 'green' sheet metallization include the following.

- Use a vacuum hold-down fixture. This is essential since the unfired tapes are flexible and thin.
- Use a multiple hole alignment fixture. Have more than one set of pins and holes since the holes can stretch and distort during use. It is suggested that different sets of pins be used for via filling, screen printing, lamination and blanking.
- A 200 mesh 304 stainless steel screen fabric is recommended.
- When screen printing green sheets rotate alternate layers 90° from the previous layer. This would lead to a stacked structure in the multilayer where alternate layers are either oriented in the casting or cross-casting direction. It has been observed that this leads to more uniform shrinkage of the sintered piece since it tends to balance out the shrinkage anisotropy observed in the casting and cross casting direction.

Some of the advantages of 'green' sheet metallization include the following.

- The conductor width and spacing shrink with the substrate during cosintering. This can yield very fine lines which are close together. Lines and spaces as small as 50 μm can be produced.
- Line spreading is virtually eliminated by solvent penetration into the substrate which increases the paste viscosity.
- Processing can be done 'reel to reel'. A spool of punched tape can be metallized in line with a blanking punch at the exit end. The use of an air flotation, low tension tape transfer system is usually used in these processes.

(ii) Lamination
The procedure for the lamination of 'green' tape cast and metallized ceramic sheets was patented in 1965 by W. J. Gyurk of RCA [22].

All of the basic principles for lamination were described in that patent. The parameters which must be controlled are temperature, pressure and time. The temperature of lamination is dependent upon the organic phases present in the tapes, i.e. the binder and plasticizers. The type of binder and the final T_g obtained through the plasticizer addition will determine the maximum and minimum lamination temperature which can be used. The number of layers and the final thickness of the tape stack to be laminated also must be considered. The time at temperature must be sufficient to heat the entire stack of tape layers and any parting layers such as Mylar which may be used to prevent sticking of the tape to the heated plattens. Sufficient pressure must be applied to provide intimate contact between the layers without distorting the laminated part and changing the final dimensions. The procedure used in typical green-sheet lamination is as follows.

1. Stack the metallized parts in the exact order desired using alignment pins.

2. Use a parting substrate such as silicone-coated Mylar on top and under the stack. Usually 0.076 mm thickness or less is desired.
3. Apply a low pressure and allow the stack to preheat for a long enough period to reach equilibrium. This can range from seconds to minutes depending upon the number of layers and the thickness of the stack. Temperatures used range from room temperature to 100 °C depending upon the organics.
4. Apply the full lamination pressure. The pressures used range from 1.4 to 138 MPa.
5. Hold pressure for times ranging from seconds to minutes. Usually about 3–4 min is adequate for complete lamination.

There are changes and innovations to the basic procedure as outlined above. All of these have been added in an attempt to improve the lamination process. The most common problem observed with heated platten lamination is the uneven application of pressure to the stack due to slight misalignment of the plattens. It is very difficult to align and to keep aligned plattens which are subjected to high pressure and heat. Uneven pressure application during lamination causes distortion during sintering due to uneven shrinkage of the part. One of the 'tricks' which is used to minimize this effect is multiple laminations of the same stack with rotation of the stack between pressings.

Another technique which was described recently [20] involves the application of isostatic pressure to the stack. The result is that equal pressure is applied on all surfaces of the multilayered stack. The author claims that package distortion is virtually eliminated by using this procedure.

(iii) Blanking to final shape

The final step in multilayer part fabrication before sintering is the removal of the alignment hole section and cutting the part to its final shape. This is usually done in a tool set made up of a punch and die. The process actually blanks the entire part in one punching step. Complicated shapes can be generated using this technique. Another technique is to score the shapes into the top surface of the package and then to snap off the edges after sintering.

One can also dice the final shape after sintering using a diamond blade cutoff saw. This is slow and creates a large amount of debris on the package. It is also not very easy to produce odd-shaped parts.

(b) Screened multilayer ceramics

In this process, which was developed by the Hitachi Corporation in the early 1970s [23] the multilayer package is built up on a single piece of unfired tape cast ceramic. The conductor and dielectric patterns are screen printed in alternate layers on the substrate. This sequence can be repeated as many times as necessary to produce the layers required. The work reported by Hitachi used alumina tapes and high temperature tungsten metallization which was cosintered at about 1600 °C. Shrinkage matching during sintering is essential for all components: substrate, dielectric and metallization. Layer to layer metallization is provided by screening metal paste through openings left in the screened dielectric layers.

Low temperature substrates, dielectrics and metallization can be used in a screened multilayer package as described above.

REFERENCES

1. Ragan, R.C. (1961) *Method for continuous manufacture of ceramic sheets*. US patent 3, 007, 222.
2. Ragan, R.C. (1963) *Method for continuous manufacture of ceramic sheets*. US patent 3, 097, 929.
3. Howatt, G.N., Breckenridge, R.G., Brownlow, J.M. (1947) Fabrication of thin ceramic sheets for capacitors. *J. Am. Ceram. Soc.*, **30**, 237–42.
4. Howatt, G.N. (1952) *Method of producing high-dielectric high-insulation ceramic plates*. US patent 2, 582, 993.
5. Park, Jr., J.L. (1961) *Manufacture of ceramics*. US patent 2, 966, 719.
6. Mistler, R.E. (1990) Tape casting: the basic process for meeting the needs of the electronics industry. *Am. Ceram. Soc. Bull.*, **69**(6), 1022–6.
7. Stetson, H.W. (1965) *Method of making multilayer circuits*. US patent 3, 189, 978.
8. Mikeska, K., Cannon, W.R. (1984) Dispersants for tape casting pure barium titanate, in *Advances in Ceramics, Vol. 9, Forming of Ceramics* (eds J.A. Mangels and G.L. Messing), American Ceramic Society, Columbus, OH, pp. 164–83.
9. Chartier, T., Streicher, E., Boch, P. (1987) Phosphate esters as dispersants for the tape casting of alumina. *Am. Ceram. Soc. Bull.*, **66** (11), 1653–55.
10. Mistler, R.E., Runk, R.B., Shanefield, D.J. (1978) Tape casting of ceramics, in *Ceramic Fabrication Processing Before Firing* (eds G.Y. Onoda and L.L. Hench), Wiley, New York, pp. 411–48.
11. Tormey, E.S. (1982) The absorption of glyceryl esters at the alumina/toluene interface. PhD Thesis, Massachusetts Institute of Technology, Cambridge, MA.
12. Shanefield, D.J., Mistler, R.E. (1976) Filter for ceramic slips. *Am. Ceram. Soc. Bull.*, **55** (2), 213.
13. Chou, Y.T., Ko, Y.T., Yan, M.F. (1987) Fluid flow model for ceramic tape casting. *J. Am. Ceram. Soc.*, **70** (10), C208–C282.
14. Runk, R.B., Andrejco, M.J. (1975) A precision tape casting machine for fabricating thin organically suspended ceramic tapes. *Am. Ceram. Soc. Bull.*, **54** (2), 199–200.
15. Grade 25025D, Silicone Coated Mylar, Custom Coating and Laminating Corp., 717 Plantation St., Worcester. MA 01605, USA.
16. Piazza, J.R., Steele, T.G. (1972) Positional deviations of preformed holes in substrates. *Am. Ceram. Soc. Bull.*, **51** (6), 516–18.
17. Mistler, R.E. (1992) Tape casting, in *Engineered Materials Handbook, Vol. 4, Ceramics and Glasses* (ed. S. J. Schneider, Jr.), ASM International, Materials Park, OH, pp. 161–5.
18. Stetson, H., Schwartz, B. (1961) Laminates, new approach to ceramics metal manufacture. Part 1: Basic processes. *Am. Ceram. Soc. Bull.*, **40** (9), 584.
19. Hetherington, R.J., Melvin, G.E., Milkovich, S.A., Urfer, E.N. (1982) *Method of fabricating an improved multi-layer ceramic substrate*. US patent 4, 336, 088.
20. Zablotny, G. (1992) Improving yields in cofired ceramic packages: an examination of process and equipment. *Hybrid Circuit Technol.*, **9** (2), 33–5.

21. Pedigo, J.L., Sugden, N.J. (1992) A printer's primer. *Hybrid Circuit Technol.*, **9** (2), 28–31.
22. Gyurk, W.J. (1965) *Methods for manufacturing multilayered monolithic ceramic bodies.* US patent 3, 192, 086.
23. Ihochi, T., Otsuke, K., Maejima, H. (1972) Screened multilayer ceramics and the automatic fabrication technology, in *Proc. Semiconductor/IC Processing Prod. Conf.* NEPCON, New York, p. 71.
24. MacKinnon, R.J., Blum, J.B. (1984) Particle size distribution effects on tape casting barium titanate, in *Advances in Ceramics, Vol 9, Forming of Ceramics* (eds J.A. Mangels and G.L. Messing), American Ceramic Society, Columbus, OH, pp. 150–57.
25. Ushifusa, N., Cima, M.J. (1991) Aqueous processing of mullite containing green sheets. *J. Am. Ceram. Soc.*, **74** (10), 2443–47.

CHAPTER 6

Plastic forming of ceramics: extrusion and injection moulding

M.A. Janney

6.1 INTRODUCTION TO EXTRUSION

Extrusion is a plastic forming method. It is limited to objects of constant cross section and is best suited to objects with high symmetry such as rods, tubes, honeycomb structures and channels. Table 6.1 gives an extensive list of extruded ceramics.

Because extrusion is based on plastic forming, there exist certain mechanical requirements for extrusion to occur [1]. The first requirement is flow. The material to be extruded must be plastic (or flowable) enough during the extrusion process to form the desired cross section under the application of pressure. The second requirement is wet strength. After the material is extruded, it must be strong enough to resist deformation due to its own weight (slumping) or due to handling stresses. If either of these conditions is not met, a good extrusion will not be obtained.

6.2 FORMULATION PRINCIPLES

To achieve the desired consistency in the batch (i.e. plasticity and wet strength) one of two basic approaches may be employed. First, depend on the mechanical properties of the particulate–solvent system; Second, depend on the mechanical properties of the binder phase. As an example, consider the situation encountered when trying to extrude a coarse alumina powder, say above 5 µm in average particle size. Such a powder suspended in water or other solvent will have a short working range and will become dilatant at high volume fraction solids. The situation appears to be intractable. The first approach would modify the mechanical properties of the particulate–solvent

Ceramic Processing. Edited by R.A. Terpstra, P.P.A.C. Pex and A.H. de Vries.
Published in 1995 by Chapman & Hall, London. ISBN 0 412 59830 2

Table 6.1 Examples of extruded ceramics

Type of structure	Ceramic composition
Furnace tubes	alumina, mullite, silicon carbide, zirconia
Insulators	alumina, beryllia, steatite
Tubular capacitors	barium titanate
Brick, tile, sewer pipe	clay, shale
Catalyst supports	cordierite, alumina, silica, aluminosilicates
Resistance heaters	barium titanate
Refractories	clay, alumina, mullite, etc.
Whitewares	clay
Electronic substrates	alumina, cordierite, glass
magnets	ferrites
Kiln furniture	various
Heat-exchanger tubes	silicon carbide, mullite

system. For example, one could add boehmite ($AlOOH$), γ-alumina, bentonite clay (if the body can tolerate the silica) or other finely divided particulate phase to the body so long as it is compatible with the chemistry of the bulk ceramic. The fine particulate flocculates the larger particles by van der Waals forces and increases the degree of plasticity and lengthens the working range of the body. The pore channels in such a body are relatively large and water 'bleeds' on application of pressure. Therefore, a thickener for the solvent phase is needed. The second approach would engineer the properties of the binder phase. For example, one could produce an ultrahigh viscosity binder phase, such as is obtained in a 10wt% solution of Methocel A4M (Dow Chemical Co., Midland, MI) in water, to give a 'gel' consistency. By using this 'gel' as the vehicle for extrusion, one has produced an extrusion batch that is analogous to a highly filled polymer melt. After extrusion the strength of the 'gel' phase prevents slumping. One can go one step further and invoke the thermal gelation of methylcellulose to achieve a true gelation as is described in Corning's patent on extrusion of catalyst supports [2]. A similar approach has been used for many years in the refractories industry. However, instead of using a polymer solution as the carrier, they have used plastic clays as the carrier for large size (up to millimetre grain size) refractory grog. The clay phase provides the plasticity to form the grog into the appropriate shape.

6.2.1 Role of additives

Additives include binders, viscosity modifiers, dispersants, flocculants, and lubricants. Binders have at least two roles to play: they give the body wet and dry strength; and they thicken the solvent phase. They may also act as dispersants or flocculants, and they may act as lubricants. Typical binders are

Table 6.2 Aqueous binders for extrusion[a]

Binder type	Viscosity grade[b]	Electrochemical Type	Biodegradable
Gum arabic	VL	Anionic	Yes
Lignosulphonates	VL	Anionic	Yes
Dextrins	VL to L	Non-ionic	Yes
Poly(vinyl pyrrollidone)	VL to L	Non-ionic	No
Poly(vinyl alcohol)	VL to M	Non-ionic	No
Acrylates	VL to M	Anionic	No
Starch	L to H	Non-ionic	Yes
Poly(ethylene imine)	L to H	Cationic	No
Methylcellulose	L to H	Anionic	Yes
Polyacrylamide	L to VH	Non-ionic	No
Sodium carboxy- methylcellulose	M to VH	Anionic	Yes
Poly(ethylene oxide)	M to VH	Non-ionic	No
Alginates	M to VH	Anionic	Yes
Natural gums	H to VH	Varies	Yes

[a]Adapted from [3].
[b]Viscosity grades are defined as the concentration of binder in aqueous solution required to give a solution viscosity of 2 Pa s.

Very low	(VL)	$> 10\,wt\%$
Low	(L)	3 to 10 wt%
Medium	(M)	1 to 3 wt%
High	(H)	0.3 to 1 wt%
Very high	(VH)	$< 0.3\,wt\%$

listed in Tables 6.2 and 6.3. The most popular binders for extrusion include methylcellulose, poly(ethylene oxide), poly(vinyl alcohol), sodium carboxymethylcellulose (also known as cellulose gum), alginates, ethyl cellulose, and pitch (for graphite). For a binder to be useful in extrusion, its rheology, burnout behaviour, and green strength properties must be characterized [3–8].

The rheological characterization of polymers includes determination of viscosity grade and degree of viscoelasticity. Most binders are available in a variety of viscosity grades. The higher the viscosity grade, the less polymer that is needed to achieve a given viscosity in solution. There is an interplay between the solution properties and the green strength properties. For example, one may use 0.5wt% of a high viscosity grade or 10wt% of a low viscosity grade to achieve the same solution viscosity. However, the green strength of the part with only 0.5wt% high viscosity grade binder, will be much lower that of the 10wt% low viscosity binder. In general, the higher the viscosity grade, the higher the molecular weight of the polymer. Also, the higher the molecular weight, the greater the tendency for the polymer to exhibit viscoelastic behaviour. A high degree of elasticity in the binder phase may be advantageous if a high springback in the extrusion batch is desired, say

Table 6.3 Non-aqueous binders for extrusion

Type	Solubility	Manufacturers
Acrylics (PMMA, PMA, PBMA, PEMA)	Ketones, alcohols, glycol ethers, and some hydrocarbons	DuPont, Rohm and Haas, Polyvinyl Chemical Industries
Butyral (PVB)	Wide	Monsanto
Cellulosics (ethyl, methyl, acetate, butyrate)	Wide (some are water soluble)	Dow, Eastman Kodak, Hercules
Poly (ethylene oxide) (PEO)	Water and polar organics (can also act as a dispersant)	Union Carbide
Pyrrollidones (PVP)	Water and polar organics	GAF Corp., BASF
Styrene (PS)	Ketones and hydrocarbons	Monsanto, Dow, Mobil, Shell, BASF, ICI
Pitch	Hydrocarbons (used mainly with carbon black for graphite extrusion)	Ashland Oil Co.
Silicone resins	Not soluble (used as 'precursor binder' in silica and alumina, refractories)	General Electric Dow Corning
Waxes, oils, gums, resins	Various	Various (an excellent reference is Industrial Waxes, H. Bennet, Chemical Publishing, Inc., 1975)

to release from a central mandrel. (Springback is a term used to describe the elastic rebound of a material after is has been subjected to pressure during a forming operation. It is an important phenomenon in extrusion, dry pressing, and pressure casting. See, for example, section 4.2 in this book.) Standard texts on polymers give extensive discussion of polymer solution properties that are beyond the scope of this chapter [9–11]. For particular binders, the manufacturer's data sheets are the most reliable source of information.

Complete binder burnout is critical to the manufacture of high quality ceramics. In air there is generally no problem except with sodium carboxymethylcellulose and some natural gums; they leave a residue of inorganic contaminants such as Na, K, P, etc. There are some situations, however, where binder burnout can be a problem. An example is the use of poly(vinyl butyral) with alumina [12]. In this situation, carbon residues can be as high as thousands of ppm even after burnout in air at 700 °C for 24 h. In inert or reducing atmosphere, most binders leave some carbonaceous remnant. The exception is the acrylics, which burnout virtually clean. In recent years, there has been some work done on the use of polymer-precursor binders, which leave behind a ceramic residue. These types of binders have the advantage that they increase the effective solids loading of the batch. However, the technology does not yet exist to use these binders reliably for making ceramic parts, except for the silicone resin-bonded parts used in precision metal casting.

Another problem encountered with binders is that they are often contaminated with trash, particulate matter, and other foreign objects that routinely find their way into the bags or drums of binder powder. Also, during binder solution preparation, clumps of binder often form that do not completely dissolve in the solvent. Binder solutions, if they are used, often need to be filtered before mixing with the rest of the batch. In some binder systems, micro-organisms will grow and produce clumps of cells that need to be removed. This is especially true of natural polymers, e.g. dextrans, and some of the cellulosics, e.g. methylcellulose and its derivatives. To clean up the binder solutions, fine screens are used [13]. In the polymers industry this is often performed on a continuous basis [14].

6.2.2 Powder

The most important body parameter is the starting powder. This may appear to be a trivial statement; but, it is one that is very often forgotten during formulation. For a non-plastic ceramic powder (largely all non-clay powders), a particle size above 5 µm may present problems in batch formulation. The powder in water has a low degree of flocculation due to weak van der Waals forces. At high solids loadings this may lead to dilatancy. For powders with particle size below 1 µm, extrusion problems are generally not so severe. The attraction between the particles is higher and the permeability is lower. The degree of flocculation can be quantified by a one of several tests, including sedimentation (Fig. 6.1), filter pressing (Fig. 6.2), and wet strength (Fig. 6.3). In the sedimentation test, the higher the solids loading in the sediment, the lower the degree of flocculation; hence, RC152 in water is less flocculated than RC152 in water with additions of either oleic acid or γ-alumina. In the filter pressing and wet strength tests, the degree of flocculation is related to the slope of the line, λ, relating specific volume to the applied forming stress (filter pressing), or to the slope of the line, $m = 1/\lambda$, relating the failure shear stress to specific volume (wet strength). In both cases, the degree of flocculation increases as λ increases. In general, a flocculated system is preferred to a dispersed system if the mechanical properties of the particle–solvent system are providing the plasticity and wet strength to the system. A flocculated system has higher strength at a given solids loading and has a longer working range for systems that depend on the mechanical properties of the solid phase. A fully dispersed system may be preferred if the mechanical properties of the binder are providing the plasticity and wet strength for the system.

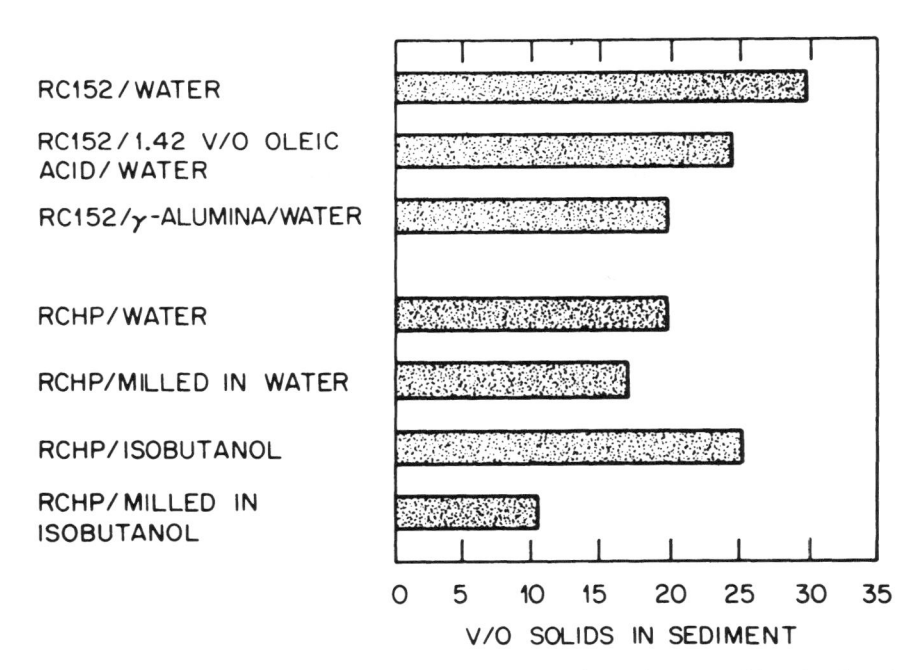

Fig. 6.1 Flocculation leads to poor packing density. The sediment packing density of a 7 vol% solids slurry increased as the degree of flocculation decreased. RC 152 and RCHP are product designations of the Reynolds Chemical Co. (Richmond, VA, USA) for alumina powders with 1.5 and 0.5 µm diameters, respectively.

6.3 MIXING

Mixing is the most critical step in extrusion batch formulation. Poor mixing will haunt the processing engineer during forming, handling, drying and firing, and in the final part. Perhaps the biggest mistake that is made in extrusion batch formulation is to try to modify an existing procedure to accommodate extrusion. One must begin with extrusion in mind. In situations where it was attempted to engineer around an existing process, the rationale was always 'to save money and processing steps'. In every case, it was discovered (often after investing 6–24 months and countless man-hours) that it was necessary to go back and **start from scratch.**

The fundamental purpose of mixing is to produce a uniform extrusion batch. With this in mind, one might suppose that a dispersant would be helpful. However, there is a conflict: for extrusion, one typically wants as much structure in the particulate phase as is possible and a dispersant destroys structure.

There are several approaches that can be taken to mixing for extrusion batch preparation including brute force, filter pressing, and some form of slurry mixing followed by drying.

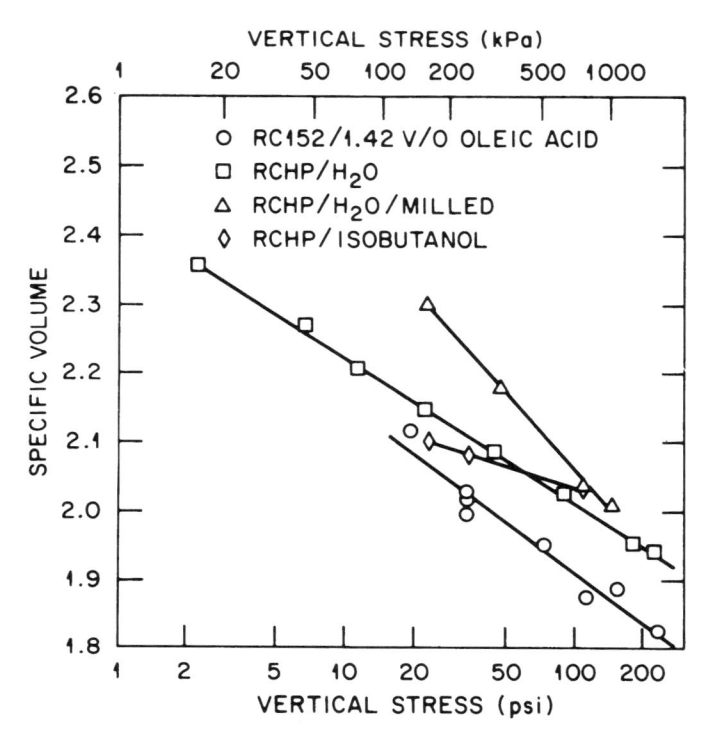

Fig. 6.2 A semilog relationship exists between specific volume (defined as $1/\rho_{\text{relative}}$) and the applied pressure for filter pressing of most materials. The slope of the consolidation line, λ, is a measure of flocculation. As λ increases, the degree of flocculation also increases.

The brute force approach typically entails the use of a high intensity mixer such as a sigma blade or muller mixer or a pug mill. The dry powder and binder solution are loaded into the mixer and the extreme mechanical action forces the powder and binder together. The situation is analogous to mixing honey and flour. Wet-out of the powder by the binder solution can be a problem. Furthermore, incomplete mixing is common. Pockets of binder and powder form inhomogeneities, which become voids in the final fired part. Poor mixing often leads to dilatant flow behaviour. This occurs because of occluded binder that resides within agglomerates. This binder is effectively not available to help with the flow of the system. Therefore, the effective solids loading of the system is higher than the actual solids loading. This type of dilatancy will disappear if better mixing is achieved. Contamination of the batch with metal impurities caused by wear of the mixer is often a problem in this approach. To mitigate this problem, the mixing blades and inner surfaces of the mixer are coated with wear resistant materials such as hard facing

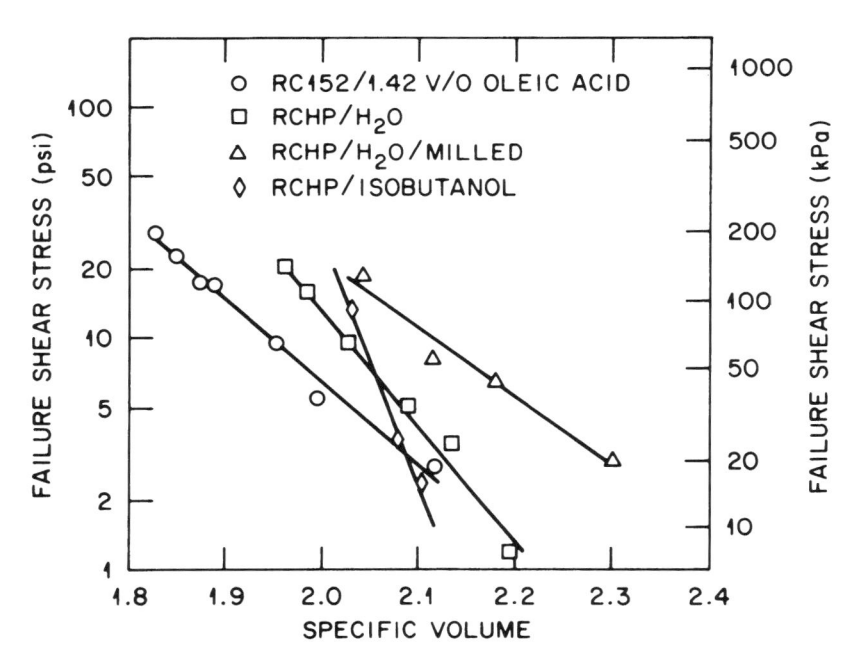

Fig. 6.3 Shear strength of wet, filter pressed bodies, τ_f, increases with decreasing specific volume, v, in a semilog manner. The slope of the curve, m, is equal to $1/\lambda$ the slope of the filter pressing curve for that material, see Table 6.5.

(stellite, WC or TiB$_2$) or polymer coatings. The brute force process can be modified to accommodate some of these problems. For example, the powder and the solvent alone can be added to the mixer. The lower viscosity liquid phase wets the powder faster and better mixing of the powder and the solvent is achieved. After initial mixing is accomplished, the binder powder is added and mixing is continued to dissolve the binder.

Filter pressing is typically used in the whitewares industry for making plastic bats that are used in extrusion and other plastic forming operations such as jiggering. Clay and water are mixed at 20–30 vol% solids. The slurry may even be ball milled if a very high degree of homogeneity is required. The slurry is then filter pressed to a high solids loading. This approach is only viable for low viscosity vehicles such as water. Clay bodies are routinely filter pressed because they have good plasticity simply on addition of water and no polymer solutions are necessary.

The third approach is based on drying of a fluid suspension. In this case the powder and solvent are mixed at a fairly low solids level, say 20 vol% solids, to effect a good dispersion of the powder and the solvent. Initial mixing can be accomplished in a high intensity slurry mixer or even in a ball mill. After initial mixing, the slurry is transferred to a high intensity mixer. Binder powder is added and mixed in to dissolve the binder. Heat and/or vacuum are then applied to drive off the excess water and to raise the solids loading of the extrusion batch. When the appropriate solids loading is achieved, drying is stopped.

At this point it is appropriate to discuss shrinkage as it relates to batch formulation. In extrusion, the solids loading of the extrusion batch determines the total shrinkage that will occur. The solids loading that is formulated is what you must use. One cannot simply increase or decrease the forming pressure as is done in dry pressing if the shrinkage is wrong. Shrinkage can be defined in several ways [15]; the concept of shrinkage factor is used in this discussion. Shrinkage factor, SF, is defined as:

$$SF = \text{green dimension/fired dimension.} \tag{6.1}$$

To relate this to density change requires a little algebra. For a cube, the volume change on firing is given by

$$V_g/V_f = L_g^3/L_f^3, \tag{6.2}$$

where V = volume, L = length, and subscripts g and f stand for green and fired, respectively. The density change is given by

$$(W_g/V_g)/(W_f/V_f) = \rho_g/\rho_f \tag{6.3}$$

where W = weight and ρ = density. For an initial approximation, assume that $W_g = W_f$, and that $\rho_f = \rho_{\text{theoretical}}$ (i.e. the fired part is fully dense). Combining equations (6.1), (6.2) and (6.3) gives:

$$(SF)^3 = (L_g^3/L_f^3) = (\rho_f/\rho_g) = 100/\rho_{g,\text{rel}} \tag{6.4}$$

Table 6.4 Typical shrinkage factors and relative green density

Shrinkage factor (SF)	Relative green density ($\rho_{g,\text{rel}}$)
1.30	45.5
1.28	47.7
1.26	50.0
1.24	52.4
1.22	55.1
1.20	57.9
1.19	59.3
1.18	60.9
1.17	62.4
1.16	64.1

where $\rho_{g,rel}$ is the relative green density. Typical values for SF and relative density are given in Table 6.4.

6.4 SOIL MECHANICS

Soil mechanics has been applied successfully to describe the mechanical behaviour of ceramic particulate systems [16]. Applications have been successful in both characterization and testing of powders, and modelling of forming processes, such as extrusion. Specifically, it has been shown that soil mechanics could be used to predict the variation of extrusion pressure with volume fraction solids and that it could also predict the variation of wet strength in filter pressed bodies with forming pressure and solids loading. An additional example of the application of continuum mechanics to forming of ceramics is given in Chapter 4.

6.4.1 Extrusion theory

Goodson [17] showed that for a wide range of clays and clayware bodies there was a relationship between the extrusion pressure, P, and the water content, w

$$P = A \exp(-\alpha w). \tag{6.5}$$

where A and α are materials parameters dependent on the particular clay or clayware body under test. However, no systematic approach was developed to describe how A and α were affected by body parameters. In 1971, Arakawa *et al.* [18] showed that the parameter α was related to the degree of flocculation of the clay body. This was done by systematically varying the degree of dispersion of a single clay body, then measuring the flow curve of that body in an auger-type extruder. In 1982, Janney [19] derived a similar equation based on theoretical soil mechanics. The approach was based on critical-state soil mechanics [20, 21]. Critical-state soil mechanics treats saturated soil (or other saturated particulate systems) as a strain-hardening, elastic–plastic material. The properties of the material depend on both its compaction history (density) and its strain history. Using this approach, an expression for the variation of yield stress with water content was derived

$$\sigma_v(\text{yield}) = C \exp[-(\rho_s/\rho_l)(w/\lambda)]. \tag{6.6}$$

where C is a materials constant, ρ_s is the solid density, ρ_l is the liquid density, and λ is a measure of flocculation. Note that the form of the equations (6.9) and (6.10) is similar. It has been shown [19] that the materials constants (A and C) in the two equations are identical and further that the two measures of flocculation are functionally related to one another

$$\alpha \equiv (\rho_s/\rho_l)(1/\lambda) \tag{6.7}$$

Thus, by applying the critical state soil mechanics formalism to the process of extrusion, a predictive model for the variation of extrusion stress with water content and degree of flocculation was attained that is identical to established empiricism for clay-based systems.

6.4.2 Wet strength

Soil mechanics can also be used to describe the variation of wet strength, τ_f, with water content, w (or specific volume, $v = 1 + w(\rho_s/\rho_l)$), and consolidation pressure, P_c, in ceramic bodies. Thus,

$$\ln \tau_f = \ln(M/2) + (\Gamma/\lambda) - (v/\lambda) \tag{6.8}$$

Table 6.5 Excellent agreement is observed between the inverse of consolidation curve slopes, λ, and $\ln \tau_f - v$ slopes, m, for various alumina compositions [16]

Material	$1/\lambda$	m
RC152/1.42 vol% oleic acid/water	-8.8	-8.2
RCHP/water	-11.1	-10.9
RCHP milled in water	-6.2	-6.5
RCHP/isobutanol	-22.0	-24.5

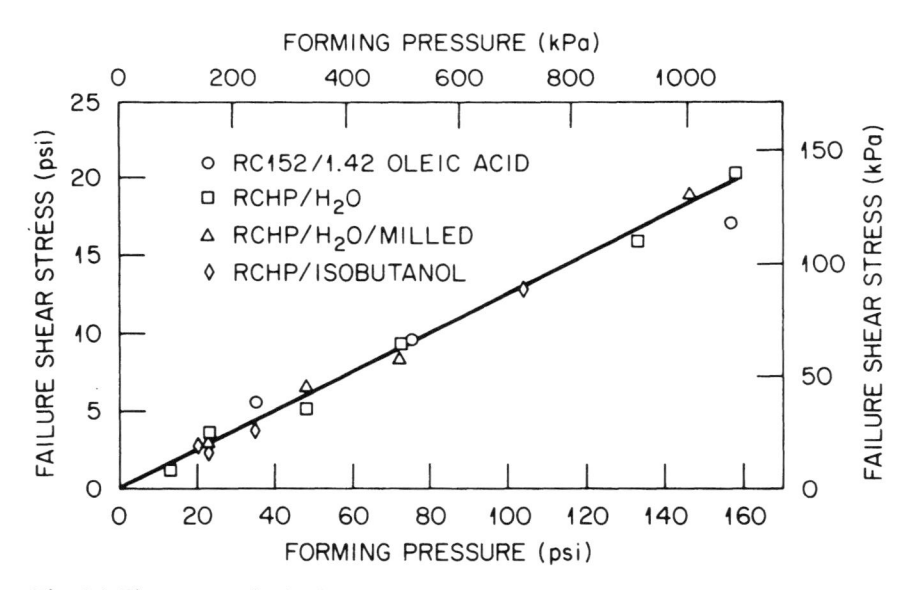

Fig. 6.4 There was a single, linear relationship between the failure shear stress and the forming pressure for all of the alumina systems tested. The slope of the line was 0.125.

and

$$\tau_f = \tfrac{1}{2} M \exp(1 - \kappa/\lambda) P_c \tag{6.9}$$

where M, κ and Γ are materials constants. Figure 6.3 shows the shear strength–specific volume relationships for samples formed by filter pressing in four ceramic systems with different degrees of flocculation. All four systems exhibit behaviour in accord with equation (6.12). This behaviour is consistent with that observed by Russell and Hanks [22] for clay bodies. Table 6.5 shows a comparison of the slopes of the consolidation curves, λ, for these four bodies and the slopes of the $\ln \tau_f - v$ curves, m, from Fig. 6.3. Excellent agreement between m and $-1/\lambda$ was obtained for all four cases, as predicted by equation (6.12). Figure 6.4 shows the strength–consolidation pressure behaviour for the same four systems. A single line describing all four systems was obtained. This result is consistent with the predictions of equation (6.13). For a given material system, such as alumina, the factor $M \exp(1 - \kappa/\lambda)$ is typically a constant [21].

6.5 THE EXTRUSION PROCESS

6.5.1 Equipment

An extruder is a machine that forces material (clay or other ceramic mixture) through a die by applying pressure. There are two basic types: the piston and the auger.

The piston extruder is very simple in construction. It consists of a barrel, a piston, and a die (Fig. 6.5). There is a minimal amount of wear (and hence minimal amount of contamination) because the design inherently minimizes the amount of contact between the extrusion mix and the extruder. It is inherently a batch process; after the extrusion mix is loaded into the barrel, no more can be added until the extrusion press run is finished. A limited amount of mixing occurs in the extruder. This occurs mainly during the reduction of the extrudate from the barrel section to the die section. Generally, few defects are generated in the extruder. The defects that are generated are typically associated with flow of material around the 'spider' or bridge used to make holes in the part being extruded. The piston extruder is typically used for making technical ceramics and for pilot or lab trials.

In contrast with the piston extruder, the auger extruder is a complex and complicated piece of equipment (Fig. 6.6). Wear in the auger extruder can be serious; it can drastically reduce the lifetime of the auger and barrel and it contaminates the product with metallic particles. Auger extrusion is inherently a continuous process. It can incorporate a mixing stage with the forming stage. Typically, there is a pug mill or other mixer such as a sigma-blade mixer associated with the extruder (Fig. 6.6). The potential capacities of an auger extruder are enormous and can be as high as 100 tonnes h^{-1}. Auger extruders are routinely used in the heavy clay, refractories, and whitewares industries. In

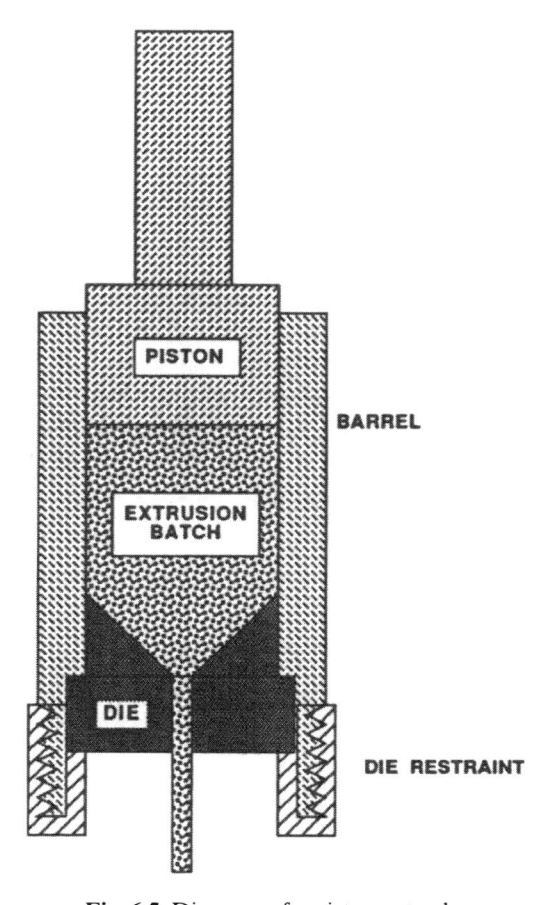

Fig. 6.5 Diagram of a piston extruder.

addition, they are sometimes used in technical ceramics applications such as capacitors and thermocouple tubing.

6.5.2   Mechanics of flow

(a)   Plastic bodies

Extrusion bodies in ceramics are almost always described by a plastic rheological model. By this is meant that the flow curve for the body (a graph of shear stress against strain rate) shows a non-zero intercept on the stress axis. Figure 6.7 shows the four plastic rheological models: perfect plastic, Bingham plastic, shear-thinning plastic and dilatant plastic. The perfect plastic body is purely a theoretical construct and is useful only for describing the behaviour of a material at vanishingly small strain rates. The Bingham plastic body is

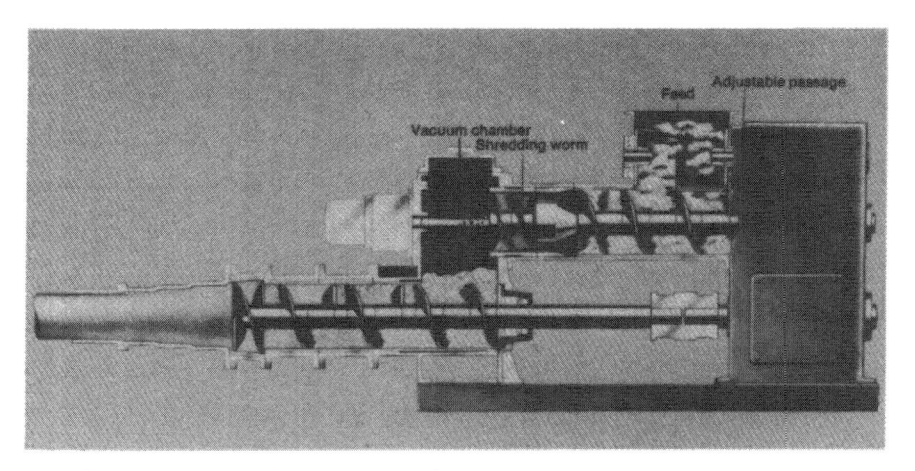

Fig. 6.6 Diagram of an auger extruder. Courtesy of Netzch and Co., GmbH.

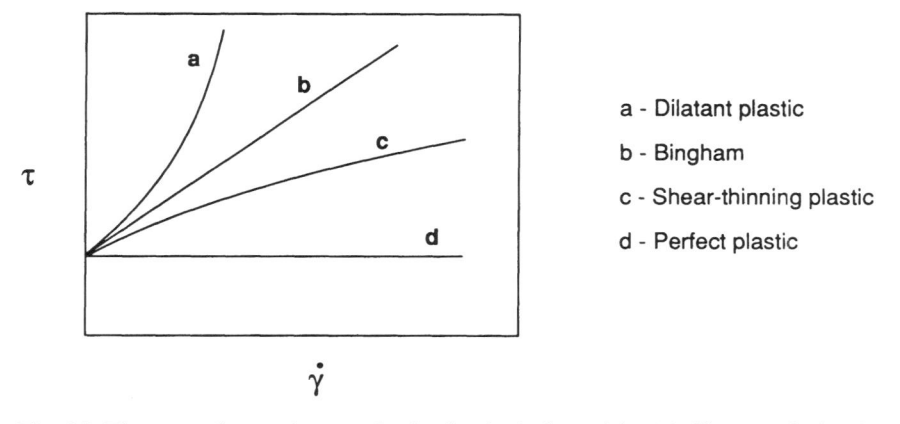

a - Dilatant plastic

b - Bingham

c - Shear-thinning plastic

d - Perfect plastic

Fig. 6.7 There are four primary plastic rheological models: (a) dilatant plasic, (b) Bingham plastic, (c) shear-thinning plastic, and (d) perfect plastic.

widely observed experimentally, and provides a useful starting point for many theoretical analyses. The shear thinning and dilatant plastic bodies are refinements on the Bingham model. They are observed commonly as well. The shear-thinning plastic body is often observed for bodies in which the particles are anisotropic or in which there is a high viscosity binder phase. In the case of the anisotropic particles, the higher the degree of shear, the greater the alignment of the particles and the lower the resistance to flow of the body. For the high viscosity binder system, there will be a strong decrease in the viscosity of the binder solution with increasing shear rate. The dilatant plastic body is universally observed at high solids loading. It is also commonly observed in systems that are poorly mixed. In such cases, the binder phase is occluded in soft agglomerates and is not available for flow. This makes the 'effective' solids loading of the body much higher than the actual solids loading.

(b)   *Plastic flow*

One of the characteristics of a plastic body is the phenomenon of plug flow. Figure 6.8 shows schematically the flow of a plastic body through a cylindrical die. Note that the central region of the material experiences no plastic flow; the shear stresses are below the yield stress, τ_y. Often the case is even more extreme; in many systems there is slip at the wall of the die and the entire column of material being extruded moves as a plug. This is especially true when external lubrication is applied to the walls of the die as is the case in brick and refractory extrusion.

(c)   *Pressure variations in the die and the barrel*

As the material being extruded moves from the barrel of the extruder through the die, a pressure profile is established. Figure 6.9 shows that there are three

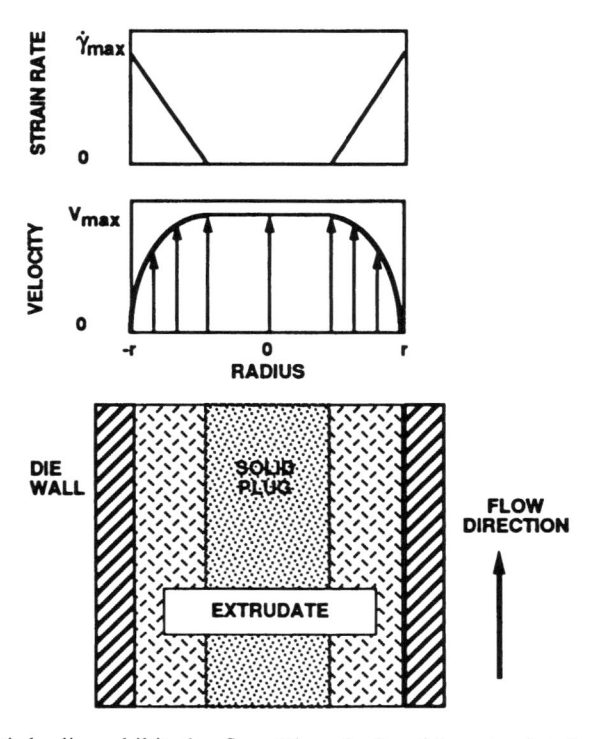

Fig. 6.8 Plastic bodies exhibit plug flow. The velocity of the extrudate is greatest in the centre. The strain rate in the extrudate is greatest at the wall and zero in the central plug.

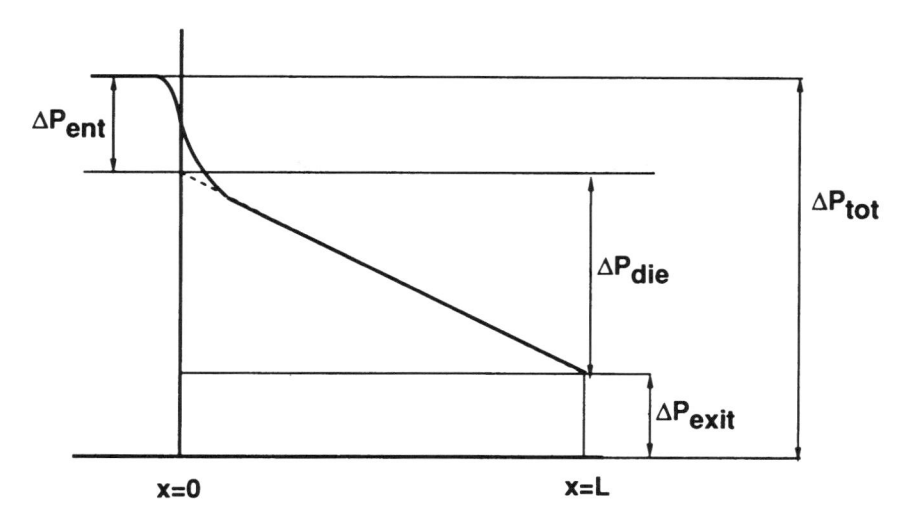

Fig. 6.9 The pressure profile in the barrel and the die consists of large pressure drops at the entrance to the die and along the length of the die, with a smaller pressure drop at the exit of the die.

distinct components to the total pressure drop. The entrance pressure drop, ΔP_{ent}, reflects the amount of work required to deform the material from the diameter of the barrel to the diameter of the die. ΔP_{die}, reflects the work required to overcome the frictional force of making the material slide through the die section. ΔP_{exit} reflects the release of pressure as the material exits from the die. In practice, ΔP_{ent} and ΔP_{die} are large and of similar magnitude, while ΔP_{exit} tends to be relatively small. However, ΔP_{exit} can still be extremely important. If ΔP_{exit} is too large, there will be a significant amount of 'springback' or elastic rebound of the extrudate. Excessive springback can lead to axial cracking and warpage of the material after it is extruded from the die. In general, ΔP_{exit} increases with increasing extrusion rate. Therefore, cracking can sometimes be eliminated by extruding at a lower rate. ΔP_{exit} also decreases as the die is made longer [23]. This was noted in the ceramics literature as early as 1949 by Mosthaf [24] who showed that cracking could be drastically reduced by using longer die sections, especially in rectangular cross section dies. Another method that is used to reduce cracking caused by high ΔP_{exit} is to incorporate a short relief section at the end of the die. The relief section consists of a short negative taper section that allows the extrudate to expand slowly over the last 1–5 cm of the die instead of expanding abruptly at the very end of the die.

The pressure profile is actually somewhat more complicated because there is drag of material at the cylinder wall as it is forced through the barrel. This leads to a falling pressure in the barrel as extrusion proceeds (Fig. 6.10). If the method of inverted extrusion is employed, then a constant pressure in the barrel is obtained. (Inverted extrusion uses a moving die with a stationary

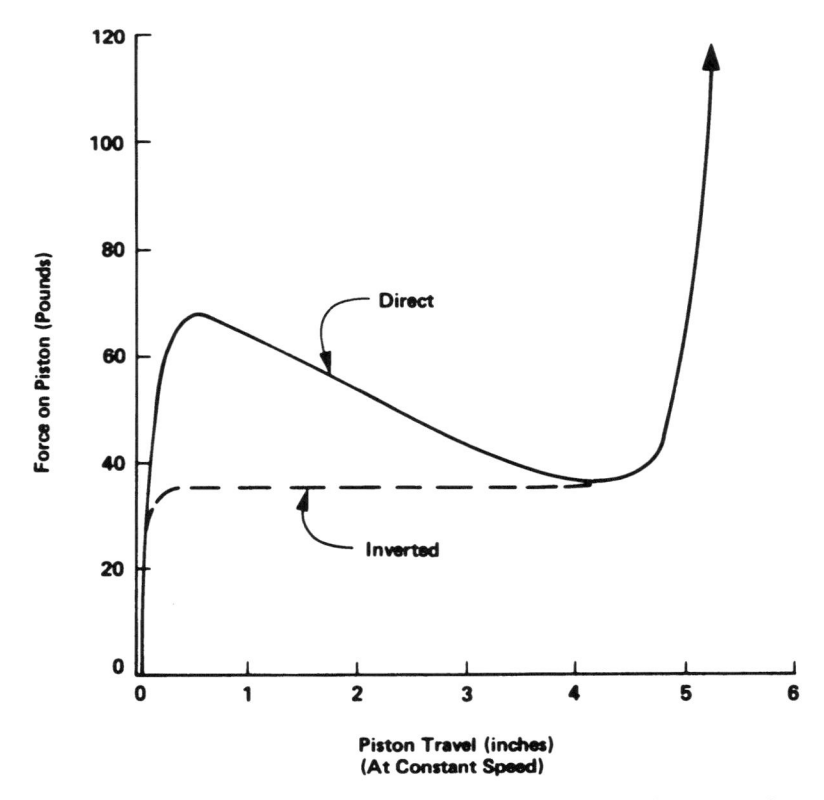

Fig. 6.10 Drag of material along the barrel wall contributes significantly to the total pressure drop. Drag at the wall is eliminated by inverted extrusion.

piston; thus, the material in the barrel of the extruder remains stationary with respect to the barrel of the extruder.)

6.5.3   Dies

The die is where the action is in extrusion. A well designed, complex die will take a long time to design, test and get into production. In designing a die, do not neglect the experience and creativity of clay manufacturers, catalyst manufacturers or polymer processors. For example, compare the complexity of a catalyst support and drain tile (Fig. 6.11) [25, 26].

Perhaps the most important aspect of designing a die is that they need to be balanced. This can be a major challenge to the die designer. In general, the pressure drop across the die is constant. Therefore, the extrusion mix will vary its speed to accommodate the geometry that it encounters as it travels through the die. Balancing even a simple tube die can take a long time to arrive at a perfectly straight piece. As the degree of sophistication of the die increases, the difficulty in balancing the die also increases.

6.5.4 Defects

The most common defects observed in ceramic extrusion are tearing, segregation, and laminations [27–31]. Tearing is a mechanical defect caused by the

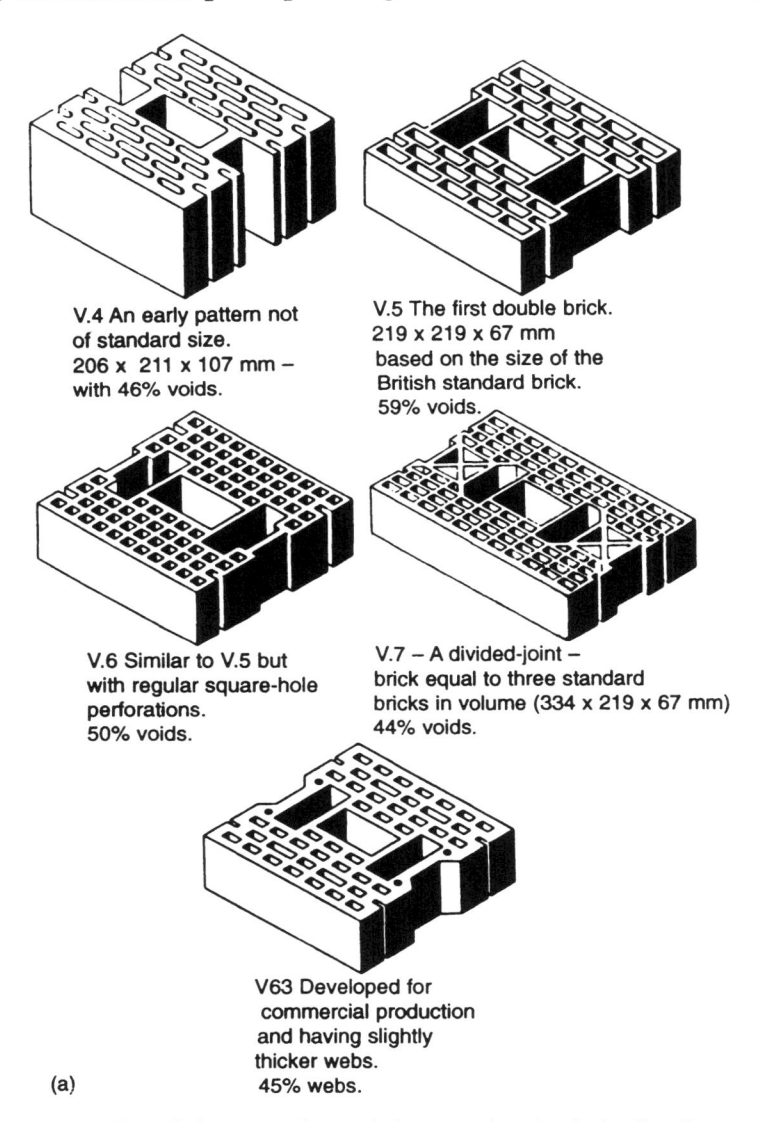

V.4 An early pattern not of standard size.
206 x 211 x 107 mm – with 46% voids.

V.5 The first double brick.
219 x 219 x 67 mm based on the size of the British standard brick.
59% voids.

V.6 Similar to V.5 but with regular square-hole perforations.
50% voids.

V.7 – A divided-joint – brick equal to three standard bricks in volume (334 x 219 x 67 mm)
44% voids.

V63 Developed for commercial production and having slightly thicker webs.
45% webs.

(a)

Fig. 6.11 Complicated shapes can be made by extrusion: (a) drain tiles; (b) (overleaf) automotive exhaust catalyst supports. Courtesy *J. Br. Ceram. Soc.* (a).

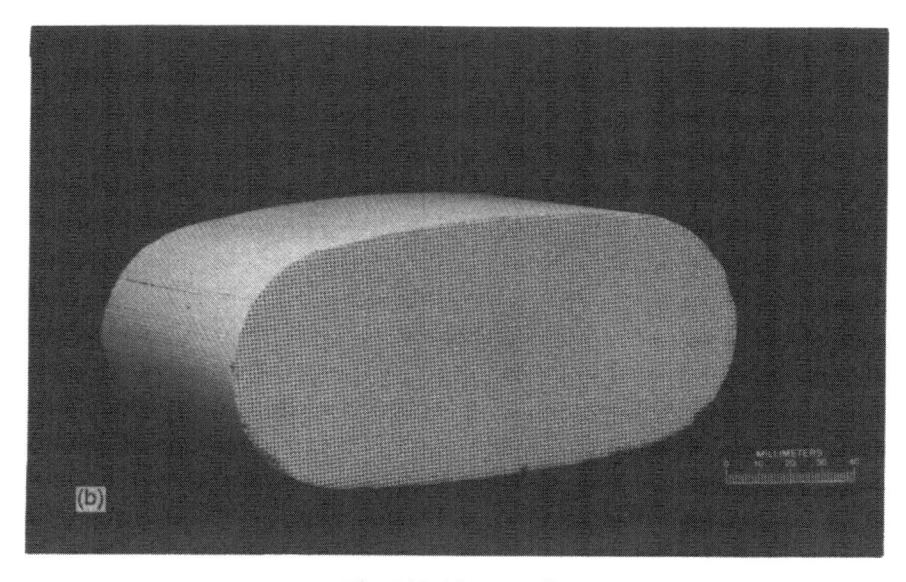

Fig. 6.11 (*Continued*)

interaction of the die with the extrusion mix. Poor die design and/or low plasticity in the extrusion mix can lead to tearing. Segregation may be caused by the differential flow of water (or other solvent) upon the application of pressure. This type of segregation is not a problem when high viscosity binder solutions are used. Segregation may also be caused by poor mixing. Pools of binder solution cause voids and laminations in the extruded part. Another form of segregation is caused by carryover of material from one batch of material to another (in piston extruders) or from early in the run to late in the run (in auger extruders). Carryover is almost always associated with poor design of the die and especially poor design of the spider or bridge used to form holes in the part. Material catches behind the spider and is only slowly released into succeeding batches of material. The water content of the trapped material and the flowing material are typically different from one another. On drying and firing, lamination cracks are formed. Lamination cracks can also form, especially in an auger extruder, because of incomplete re-knitting of the extrusion mix around the auger during its flow through the barrel of the extruder. Figure 6.12 shows typical extrusion defects.

6.6 ADDITIONAL TOPICS

6.6.1 Ultrasonic extrusion

In 1961, Tarpley [32] showed that extraordinary advantages could be achieved by using ultrasonic vibration to assist in the extrusion of ceramic

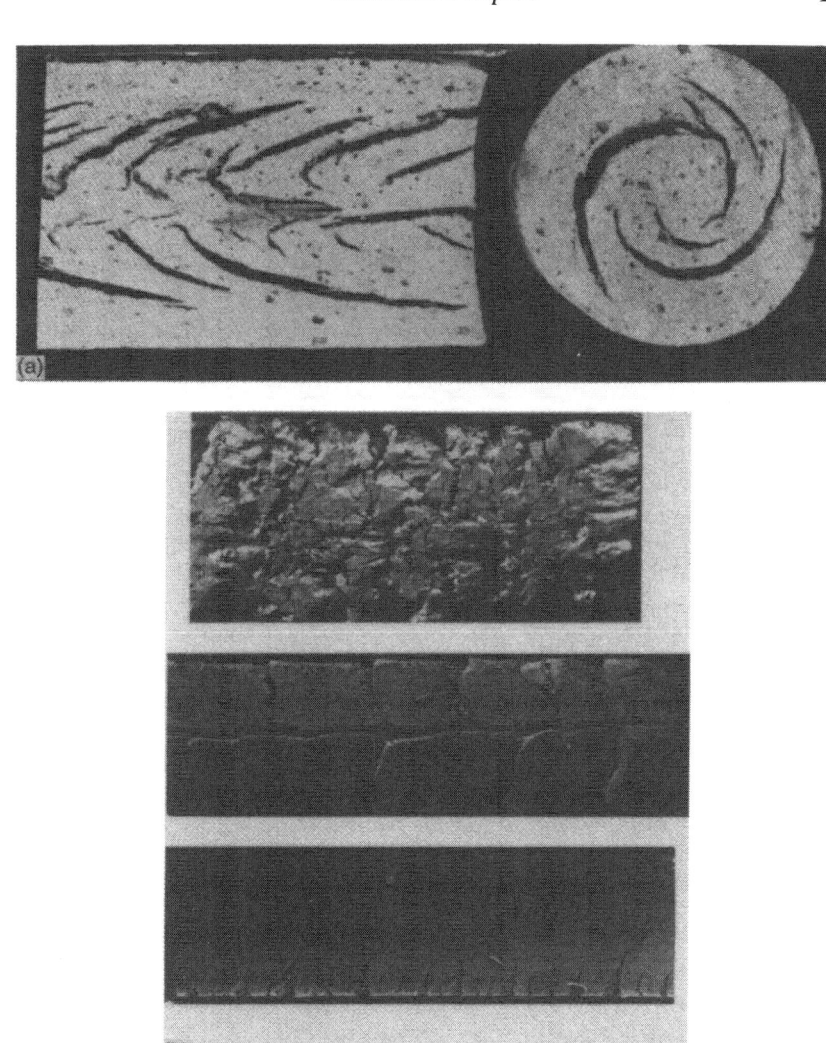

Fig. 6.12 Typical defects observed in extrusion: (a) lamination cracks in longitudinal (left) and transverse (right) section; (b) edge tearing. Courtesy G. Robinson [27].

bodies. The advantages that were demonstrated by Tarpley included reduced extrusion force, improved surface finish of the extruded part, reduced wear and pick-up of die contamination, the ability to extrude dilatant mixes, more uniform density (and therefore more uniform shrinkage), higher fired density, and higher fired strength. To the best of this author's knowledge, no one has exploited this approach commercially. Figure 6.13 shows the advantages of reduced pressure and increased rate of extrusion afforded by ultrasonic extrusion.

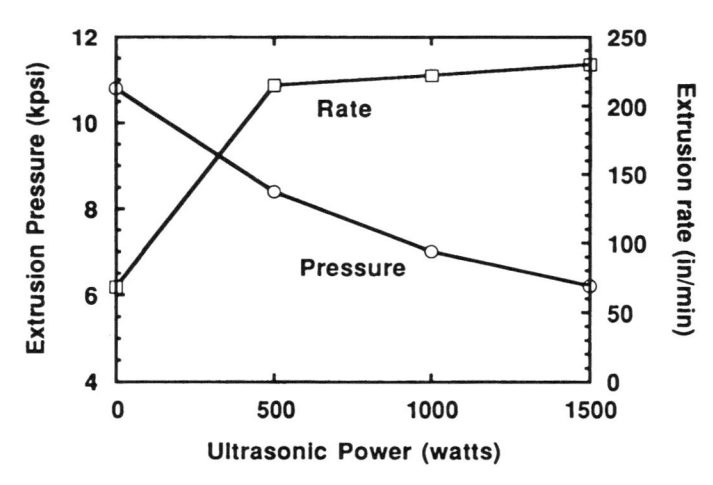

Fig. 6.13 Extrusion pressure declined rapidly with the application of ultrasonic power. Rate increased at first but soon reached the limit of the extruder. Redrawn from the data of Tarpley [32].

6.6.2 Multilayer or coaxial extrusion

Multilayer and coaxial extrusion are commonly used in the plastics industry to blend the properties of two or more materials into a single form [23]. In the ceramics industry, such an approach is rarely used. However, one notable example was found by the author and is presented here. In an article by Zivanovic and Janjic [33], an example is given coating a low grade of clay with a higher grade of clay by coextrusion. In this way the brick manufacturer was able to upgrade the quality of their bricks from common to face brick.

6.7   INTRODUCTION TO INJECTION MOULDING

Injection moulding is a method for forming complex shapes in high volume. It is currently used to mass-produce numerous small ceramic parts including cores for lost-wax metal casting, thread guides, cutting tools, welding nozzles, and other small, high production parts. It is also used on a prototype basis to produce turbocharger rotors and turbine wheels for advanced heat engine designs. Injection moulding is in general the forming method of last choice. If a part can be made by die or isostatic pressing, slip casting, or extrusion rather than by injection moulding, then it is.

When then should injection moulding be used for manufacture? Injection moulding is used when no other forming process can produce the degree of complexity necessary, when the volume of parts is high, and when the cost of the die can be justified. It is especially attractive for operations in which a lot of green machining is currently employed, e.g. thread guides. It has the potential to eliminate green density variations and the accompanying distortions on firing.

This section will cover material on injection moulding that is specific to the injection moulding of ceramic materials. Subjects that are described well and in detail in the plastics literature [34, 35] are only touched upon. Most of this section will be devoted to formulation, mixing and binder removal. Some of the newest technologies that may replace traditional injection moulding will also be reviewed.

6.7.1 A little history

Injection moulding of ceramics is not a new technology. As early as 1936, patents were appearing in the United States [36]. There are numerous other patents from the 30s, 40s, 50s, and 60s in both the US and European patent literature. The earliest open literature paper is by Karl Schwartzwalder [37]. At that time, the AC Sparkplug company was using injection moulding to make automotive sparkplugs. A thorough review of the literature has been given by Edirisinghe and Evans [38, 39].

6.7.2 Limitations

At present, injection moulding is limited to parts that have at least one dimension that is of the order of 1 cm in thickness; those applications in which the part dimensions are significantly larger than this (such as turbochargers and turbine rotors) are not commercially viable at the present time. (An exception is that some large refractories are made by injection moulding. These materials can be made in large cross sections because the particle size of the ceramic in those parts is also quite large so that binder removal is made somewhat easier.) Injection moulding will continue to be attractive mainly for high production, high value-added parts in which the cost of mould design and construction and the long binder removal times can be justified.

6.7.3 The process

The process of injection moulding a ceramic consists of four parts: mix preparation, part formation, binder removal, and firing. Preparation of the mix consists of incorporating a ceramic powder and an organic binder. This is typically accomplished in a sigma-blade or other high-intensity mixer. The molten mix typically has the consistency of toothpaste for wax-based binder systems; its consistency can be much thicker for polymer-based systems. On cooling, the mix becomes hard and brittle, and typically granulates into pea-sized pellets. Part formation begins with loading the granulated mix into the injection moulding machine, evacuating air from the mix to eliminate voids, and shooting the heated mix into a cold die. The mix fills the cavity and on cooling sets in the form of the die. The part is ejected from the die and the process is repeated. Binder removal involves the elimination of the binder (vehicle) from the part. As in many ceramic processes, the binder in injection moulding is only a temporary constituent. It is used only during part formation and must be eliminated before firing. Binder removal is typically accomplished in two steps: the bulk of the binder is removed at low temperature; and the remainder of the binder 5–10%) is removed in the early stages of the sintering cycle. Firing usually follows standard procedures established for similar die or isostatically pressed parts.

One of the special considerations of injection moulding is that it inherently combines the competitive processes of heat transfer and fluid flow. White [40] has described injection moulding as a 'non-isothermal, fluid mechanical process involving solidifying liquids with complex rheological properties'. In the long run, this combination of transport mechanisms leads to a host of problems, including knit lines, sink marks, short shots, excessive shrinkage, flashing, and jetting [35]. It would be desirable to separate the two transport processes from one another; i.e. to separate the mould filling step from the setting step. However, few of the commercially used processes do this.

6.8   DETAILED DISCUSSION OF THE PROCESS

6.8.1   Powder considerations

In general, the smaller the particle size, the higher the viscosity of the injection moulding mix. Broad or multimodal distributions are preferred to narrow

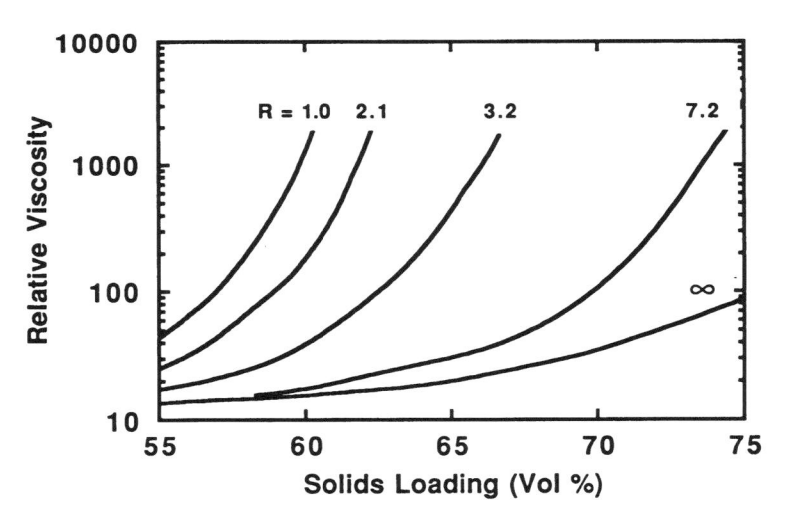

Fig. 6.14 For a given solids loading, the viscosity of a dispersion of particles in a liquid decreases as the ratio, R, of large to small particle size increases. The figure shows R values from 1 (monomodal) to nearly infinity.

distributions to reduce the viscosity of the mix (Fig. 6.14). The coarser the powder, the higher the viscosity of the binder needed to achieve a good injection moulding mix. At least two reasons can be cited for this: enhanced settling and higher permeability. As the particle size of the powder increases (especially above 5 μm), there is a greater tendency for the particles to settle due to gravitational forces. Also, with increasing particle size comes increasing pore size; hence the permeability of the powder bed increases. To prevent migration of the binder during the application of pressure, the viscosity of the binder must be increased.

As in any ceramic process, the existence of agglomerates in an injection moulding mix is bad. Soft agglomerates create problems in mixing. Wet mixing using a solvent that will dissolve the binder and other processing aids (either with or without ball milling) is very effective in eliminating soft agglomerates from the batch. This typically requires the additional step of drying the batch down to just the powder and binder; but it is well worth the extra effort. Hard agglomerates create problems in every stage of processing and they can be the site of strength-limiting flaws. A powder that is free of agglomerates is essential to making a good injection moulding mix.

Finally, it is essential that the powder that is used for injection moulding be prepared specifically for that purpose. One cannot make the powder from another process line (e.g. dry pressing) work in an injection moulding line. It is imperative that one starts from scratch.

6.8.2 Formulation

If a good formulation can be developed, then good injection moulded parts can be made. The key is to develop the proper formulation for a particular powder and application. Formulation typically means picking a binder and dispersant system for a given powder system. Binder systems can be classified into four general categories:

- thermoplastics,
- thermosets,
- sublimable organics,
- chemical gelation (e.g. ethyl silicate).

Thermoplastics have received the most attention in the ceramics community [38, 39]. Thermoplastic binders include waxes and other wax-like substances, polyethylene, polypropylene, polystyrene, ethylene–vinyl acetate copolymers, poly(butyl methacrylate) and other acrylics, and any other low-viscosity polymer systems. Thermoplastic binders typically include other additives such as oils, plasticizers, etc. Table 6.6 gives three typical thermoplastic binder compositions [41, 42].

Thermosets also have received significant attention, especially in the patent literature [38, 39]. Thermoset systems include epoxy, phenolic, urethane,

Table 6.6 Typical injection moulding binder systems

System	Constituent	Weight ratio	Role
I [41]	Isotactic polypropylene	6	Main binder
	Microcrystalline wax	2	Viscosity modifier
	Stearic acid	1	Wetting agent
II [41]	Isotactic PP	4	Main binder
	Atactic PP	4	Viscosity modifier
	Stearic acid	1	Shrinkage reduction Wetting agent
III [42]	Paraffin wax	9	Main binder
	Epoxy resin	0.5	Plasticizer
	Oleic acid	0.5	Dispersant
IV [43]	Paradichlorobenzene	10	Sublimable binder
	Ethyl cellulose ⎱ Shellac ⎰ Silicone resin	1 to 10	Auxiliary, pyrolyzable binder
V [44]	Ethyl silicate	10	Primary solvent and silica source
	Gel inducer	0.1	Precipitation aid

furfuran and furfural. Thermoset systems typically include waxes, oils, and other processing aids which aid in binder removal.

Sublimable organics provide a novel alternative to binder removal compared with thermoplastics and thermosets [43]. In a sublimable system, the main constituent of the binder is composed of an organic compound that has a high vapour pressure at temperatures below its melting point (Table 6.6). Typically the primary binder is either naphthalene or paradichlorobenzene. An auxiliary binder is added to the composition to provide green strength to the part after the sublimable binder is removed. Candidates for this auxiliary binder are shellac, ethylcellulose or another polymer that is compatible with the sublimable binder and is soluble in a common solvent. Mixing is accomplished in an organic solvent, typically trichloroethylene or other chlorinated solvent that is both non-flammable and readily evaporated at room temperature. After mixing, the solvent is evaporated from the batch, which leaves a granulated mixture of binders and ceramic powder.

Chemical gelation [44] involves precipitation of a finely-divided inorganic phase from a solvent. The inorganic phase is typically silica, although there are also systems based on alumina, zirconia and titania. The solvent phase is typically composed of a metal alkoxide such as ethyl silicate dissolved in an alcohol such as ethanol (Table 6.6). Gelation is accomplished by diffusing ammonia vapour into the body or by *in situ* base generation. The fine silica (or other material) that is precipitated bonds the ceramic particles together by van der Waals attraction to form a strong wet body.

6.8.3 Mixing

Batch mixing is most often done in a sigma or other high-intensity mixer. There are two major approaches: brute force and solvent-based. In the brute force approach, binders and other organic processing aids are mixed directly with ceramic powder at a temperature above the melting point of the binder phase. This is usually the first approach taken because it is the most straightforward. However, there are numerous pitfalls associated with this approach including incomplete mixing, residual soft agglomerates, dilatancy in the early stages of mixing, excessive wear of the mixer blades, and incomplete solution of the various binder components. An inherent limitation of this mixing route is that the size of the batch must be well-matched to the size of the mixer; otherwise, efficient mixing is not obtained. In the solvent-based approach, binders, additives, and ceramic powder are slurried together in a common solvent. The entire mix can be homogenized in the slurry state (even to the point of placing the mix in a ball mill to effect mixing). Then the mix is loaded into a mixer, heated and/or evacuated to evaporate the solvent, and then cooled to pelletize the mix. Alternatively, the mix can be pan dried, with stirring, to form the pelletized mix. The solids loading in the final mix is determined by the amount of binder and powder initially added to the solvent.

Many of the problems with poor flow of the injection moulding mix, either in the injection moulder or in the die, can be traced to poor mixing. Dilatancy is often caused by poor (incomplete) mixing.

6.8.4 Equipment

As in extrusion, there are two primary types of injection moulding machines: plunger and screw. These will be described briefly below. Detailed descriptions are available elsewhere [34, 35].

The plunger machine (Fig. 6.15) is similar in design to a piston extruder. It is a batch type machine of simple design. Contamination of the mix by the injection moulder is not a serious problem. Also, plunger machines are not as expensive as the screw machines. In fact, very low cost machines are available for initial trials [45]. The plunger machine is more adaptable, and is more suited for development work.

The screw machine (Fig. 6.16) is analogous to the auger extruder. It is continuous and is of a complicated design. Contamination and wear are serious problems because very high shear stresses are generated during flow of the mix through the machine. Screw machines are more expensive than plunger machines. A screw machine is a more 'fixed' design and is more applicable to production.

Recently, a major advance in the use of screw machines for injection moulding of ceramics has been accomplished. Pasto and Natansohn [46] have shown that the amount of metallic contamination introduced into silicon nitride parts by the injection moulding process can be greatly reduced. They

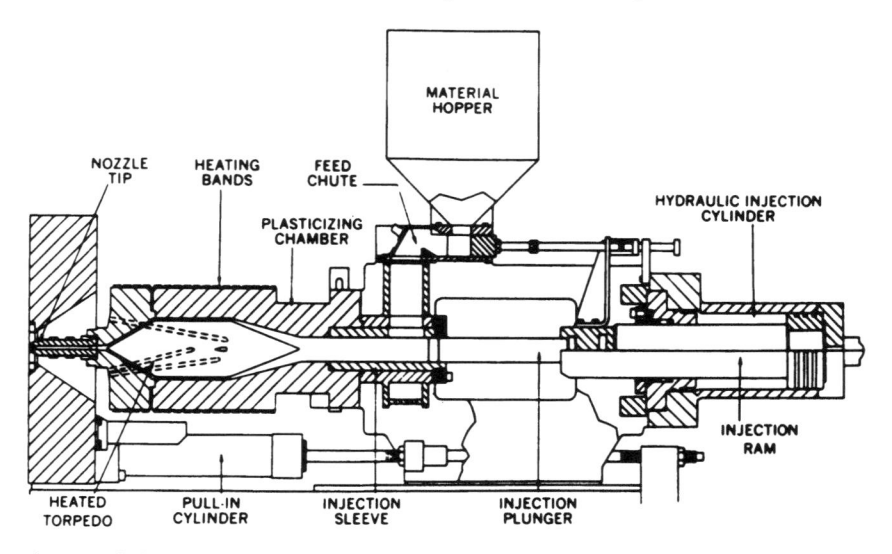

Fig. 6.15 Schematic of a plunger injection moulding machine. Courtesy of HPM Div., Koehring Co.

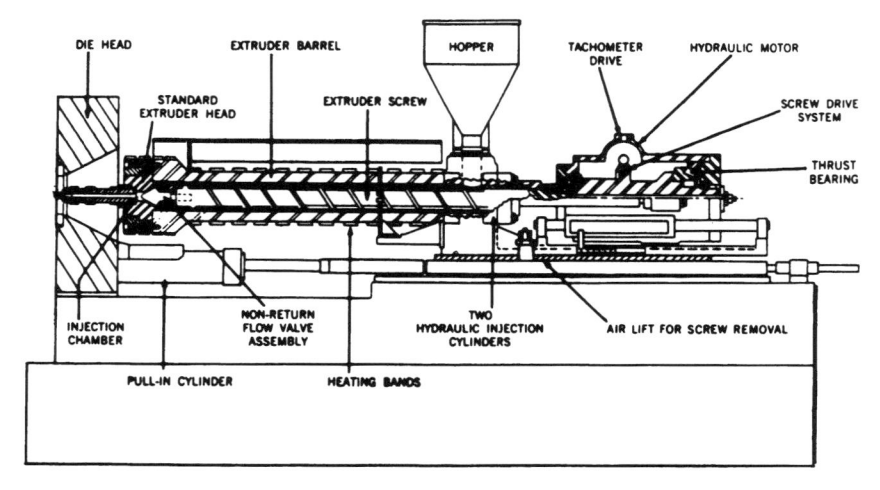

Fig. 6.16 Schematic of a screw injection moulding machine. Courtesy of HPM Div., Koehring Co.

replaced the hardened metal screws of a screw injection moulding machine with ones made of a high-performance engineering plastic, poly(ether ether ketone) (PEEK). The number and size of metallic inclusions in the injection moulded parts was significantly lower when the PEEK screws were used as compared with the metallic screws. The PEEK screws also performed well mechanically. Post-run examination of the PEEK screws revealed a small degree of polishing of the screw, but no noticeable wear in terms of galling or rounding of the screw flights.

6.8.5 Binder removal

Binder removal is the most important and the most difficult step in the injection moulding process. Parts are typically packed with a setter powder during binder removal. This is done for two reasons. First, during binder removal, the parts become very weak and soft because the binder phase melts. Therefore, the parts require mechanical support to prevent them from collapsing. If this is the only requirement of the setter powder, a coarse powder is typically used. It packs to high density and is largely unaffected by capillarity. Second, the packing powder is often selected to provide capillary suction to enhance the binder removal process. This topic has recently received much attention in the literature [47, 48]. Typical high surface area powders ($> 100 \, \mathrm{m^2 \, g^{-1}}$) used for packing include carbon black, fumed silica and hydrated alumina. Often these high surface area powders are blended with coarser powders to provide the proper mixture of capillary suction and mechanical support.

Binder removal is basically a diffusion problem and is analogous to drying. It is affected by the particle size of the powder, the packing arrangement of the powder, the viscosity and vapour pressure of the binder(s), the temperature, and the gas pressure in the binder removal chamber. It is here that many of the defects are generated, including pores, cracks, laminations, pin-holes, 'orange peel', etc. Binder removal times range from days to more than one week, depending on the particular binder system employed. Removal can be assisted using vacuum, packing powders, pressure, and oxygen, again depending on the particular binder system.

6.8.6 Moulding parameters

Many of the basic moulding parameters for ceramics are similar to those for polymers [34, 35]. However, there are many details that are specific to ceramic injection moulding; these will be covered here.

The two biggest differences between moulding polymers and ceramic mixes are elasticity and thermal conductivity. Ceramic mixes are much less elastic than their polymer counterparts and they have a much higher thermal conductivity. The lower elasticity means that ceramics rebound less after being compressed, such as when they exit the barrel of the injection moulder or when they flow through a gate before entering a mould. The higher thermal conductivity of the ceramic mixes means that they cool much faster than polymers do. Therefore, there is less time available to fill a mould than with polymers. Because of these differences, mould design for ceramics is different from mould design for polymers in a few key ways. First, in general, runners and gates in ceramic moulds are larger than in their polymer counterparts.

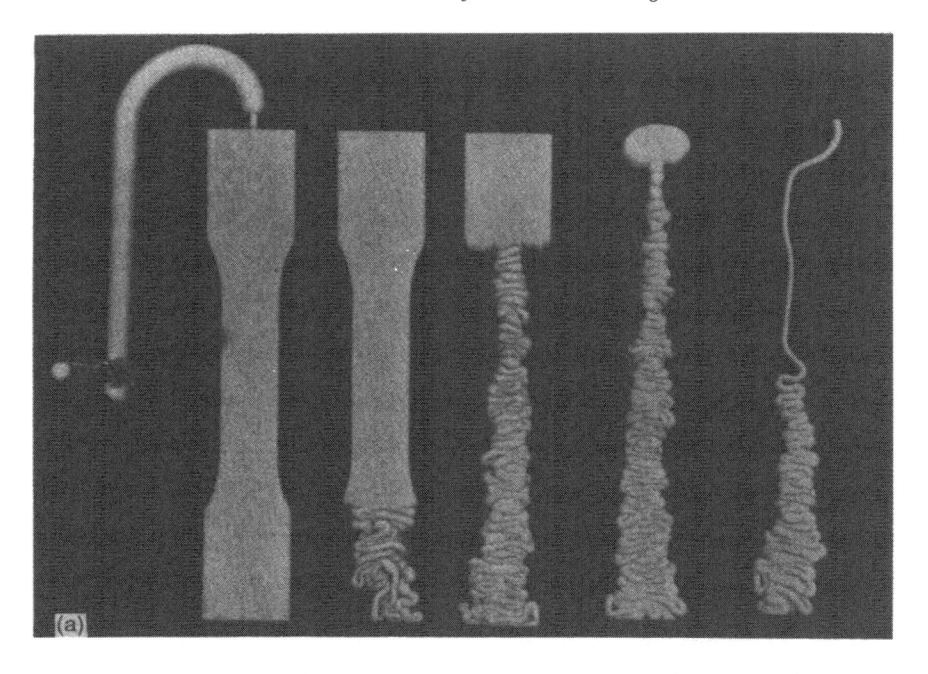

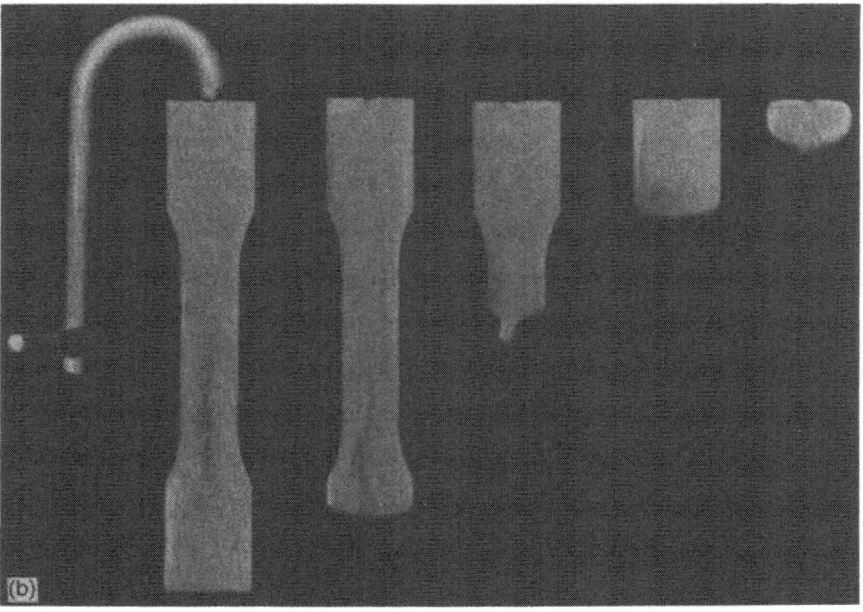

Fig. 6.17 Jetting is caused by the use of too small a gate in ceramic injection moulding. (a) A small gate leads to jetting and poor packing of material in the die. (b) A large gate fills from the gate end to the far end of the mould. Courtesy of B. Mutsuddy [49].

This is done to reduce the tendency of the runners and gates to 'freeze-off' before the mould cavity is filled. Ceramic gates are also larger than polymer gates to prevent 'jetting' of material to the far end of the mould cavity, and subsequent 'back-filling' as is shown in Fig. 6.17 [49].

It is important to match moulding parameters to the parts that are being made. This sounds trivial; however, each part will respond differently to the thermal and fluid transfer conditions encountered during moulding. In particular, parts may have to be redesigned for injection moulding. Compromises must be made between the perceived needs of the part designer and the limitations of the injection moulding process. A few of the more important design criteria are listed in Table 6.7 and others are shown in Fig. 6.18.

The time–temperature parameters used during moulding represent a compromise among throughput, materials properties and part quality. To obtain a good part, the part must be cool when is removed form the die. Otherwise it will deform during ejection or subsequent handling. The question becomes: how long should the part be allowed to cool before it is removed? But this will also depend on the temperature of the mix where it enters the mould from the injection moulder barrel. It is preferred that the mix be at as low a temperature as possible during mould filling. This increases the density of the injection moulding mix and reduces as much as possible the shrinkage in the die during cooling. However, a trade-off is encountered between good fluid flow and shrinkage. If the temperature is too high, flashing and sink-marks are encountered. If the temperature is too low, short shots, knit lines and jetting occur. The final combination of mix temperature and hold time in the mould will be a compromise that minimizes defects yet allows for a reasonable throughput of product. This can only be decided by actual moulding trials.

Table 6.7 Design guidelines for ceramic injection moulded parts[a]

Design feature	Guideline
No enclosed cavities	Cannot remove part from mould
No sharp corners	Minimum 0.2 mm radius (if possible)
Redesign thick sections	Include webs or stiffening ribs for strength, section thickness of rib should match wall thickness of main part
No sharp changes in section thickness	Taper transitions, or make multiple step changes
Locate all holes, recesses etc. symmetrically	Makes filling mould more uniform
Slots and holes should be no less than 0.1 mm	Prevents closing up of holes by flashing
Webs, ribs and other thin sections should be no thinner than necessary	Section thickness < 0.5 mm is difficult to handle without cracking.

[a]Adapted from [50].

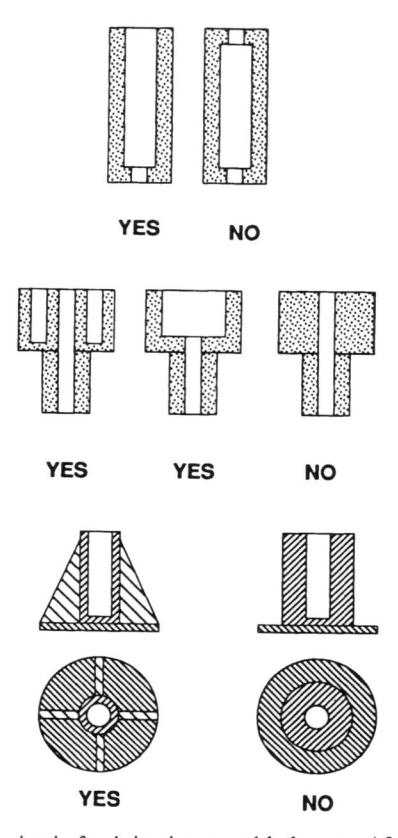

Fig. 6.18 Design criteria for injection moulded parts. After Peltsman [50].

Finally, when changing over from one material to another, it is essential that the barrel of the injection moulder be cleaned out. In very high reliability cases, complete disassembly of the barrel of the injection moulder is required. This will ensure that absolutely no material is left behind to contaminate the next batch. In other cases, it is only necessary to purge the barrel with an appropriate cleaning purge material. One approach would be to run a batch of the base wax or binder through the machine. In other cases, other appropriate purge materials may need to be used.

6.8.7 Defects

Defects may be generated during any stage of the injection moulding process. Mixing defects are especially problematic in that they will cause problems at all subsequent stages of moulding and firing. Defects introduced during moulding are somewhat less a problem, if they are noted prior to binder

Table 6.8 Defects in injection moulded ceramics[a]

Stage	Defect	Cause
Mixing	Undispersed agglomerates	Mixing not intensive enough
	Undispersed polymer	Poor temperature control in mixer
		Temperature in mixer too low
	Abrasive contamination	Poor quality dispersant
		Bad seals in mixer
		Need for lubricant in binder system
Moulding	Flashing	High mix temperature
		Pressure too high
		Mould wear
	Sink marks	Mix too hot
		Packing pressure too low
	Voids	Large – slow injection
		Small – adsorbed water
	Weld lines	Mould design
		Low mix temperature
	Cracks	Low injection pressure
		High mix temperature
		Re-entrant angles in mould
	Warpage	Part too warm at ejection
		Adhesion of part to mould
Binder removal	Deformation	Residual stresses – mould temperature too high
	Cracking	Heating rate too high
		Bloating by residual gas
		Thick/thin section problems
		Heating rate too high during high firing
	Slumping	Poor support of part by packing powder
		Insufficient packing powder to wick binder
		Heating rate too high – insufficient time to remove binder from part

[a]Adapted from [41], [51] and [52].

removal; for thermoplastic systems, at least, the material can be recycled at this point. A summary of defects and their causes is given in Table 6.8.

6.8.8 Other factors

The quality assurance tests used for injection moulded parts will often be different from those used in other forming processes. It is especially important in starting up an injection moulding line that the quality assurance group is included in technical briefings as soon as a viable demonstration of the process has been made and before production parts are identified. In addition, the specifications for both raw materials and finished parts may be different for injection moulded parts than for other processes. Make sure that these differences are understood by all involved personnel and management.

6.9 NEW APPROACHES TO FORMING COMPLEX SHAPES

In recent years, various groups in the USA and Europe have developed alternative approaches to the traditional injection moulding process for forming complex parts in mass production [53–59]. These new processes are summarized in Table 6.9. All of these processes address, in their own ways, many of the perceived shortcomings of traditional injection moulding. Their approaches are essentially independent, although not necessarily mutually exclusive. We will now review some of the inherent problems with injection moulding.

Among the limitations of current injection moulding technologies are: moulding defects, long vehicle removal times, low green strength after binder removal, warpage during binder removal, differential binder removal, and thick-section problems. Some of these difficulties are the result of generic problems related to the injection moulding process itself; others are related to the composition of the wax or polymer vehicle used to carry the ceramic particles.

During the mould filling step of injection moulding, the competitive processes of heat transfer and fluid flow work against one another [40]. It is desired that the injection moulding mix be as fluid as possible to make filling of the mould easy; this implies that the mix should be significantly superheated above the solidification point of the wax or polymer carrier vehicle. However, the moulding mix must cool to become more viscous and eventually harden to a solid to set the part in the mould. The result of this competition between the two processes is often the creation of defects in the as-moulded part. If the moulding mix cools too rapidly, knit lines, short shots, and voids occur; if it cools too slowly, flashing, jetting and sink marks occur. Thermal strains and warpage caused by differential cooling throughout the part are ubiquitous. Many of the processes listed in Table 6.9. overcome these problems by separating the mould filling operation from the setting operation.

The use of a 100% polymer- or wax-based vehicle is responsible for additional problems in injection moulding, especially with respect to binder removal and high firing. During moulding, the vehicle (wax or polymer) must be very fluid to give the injection moulding batch the high fluidity necessary to fill the mould. On cooling in the mould, the vehicle regains its rigidity and sets the part. However, during binder removal, either by capillary wicking of the polymer/wax into a packing powder [47,48] or by vapourization/oxidation [60], the vehicle must again melt and become fluid. At that point in the process, the part becomes as soft as it was during its initial injection into the mould; that is, the part has essentially zero strength and will easily warp and sag. Also, any residual gas that was trapped in the part during moulding will form bubbles or will crack and part at that point; these will become strength-limiting voids and planar flaws in the as-fired part. In addition, after binder removal, the green strength of the part is very low.

An additional problem with current injection moulding technology is the issue of making parts having thick sections, such as the hub of a turbine wheel

Table 6.9 New approaches to complex shape forming

Process	Developer	Technology
Pressure Casting [53]	Dorst Netzch Ceram Research	An amalgam of slip casting and filter pressing Depends on proprietary polymeric filter materials
Binderless injection moulding [54]	CPS, Inc.	Freeze casting of water-based slurries Solvent removed by freeze drying
Methylcellulose gelation [55]	Cabot Corp.	Reversible inverse thermal gelation of methylcellulose Methylcellulose gels on heating.
Agar gelation [56]	Allied-Signal	Reversible gelation of aqueous agarose solution on cooling Depends on hysteresis of melting and gelation
Gelcasting [57, 58]	Oak Ridge National Laboratory	Irreversible gelation of aqueous polymer gels In-situ polymerization of acrylamide and other monomers
Reversible Deflocculation [59]	Monsanto	Deflocculation provided by a degradable dispersant Reflocculation causes setup

or turbocharger. For section thicknesses over ~ 2 cm, severe problems have been reported concerning flaw generation during binder removal. To deal with these problems, several innovative approaches have been developed. GTE Labs [61] reports isopressing parts after binder removal as a means to reduce the severity of flaws in thick parts. Another approach is reported by NGK, who make rotors for Mitsubishi [62]. They make the blade section and the hub section of the rotor as two separate parts. The blade section is injection moulded and the hub section is isopressed. The two sections are joined using a slip of the same ceramic composition, much as a handle is attached to a tea cup in the whitewares industry. After firing, the part acts as a single piece, with no indication that it was formed as two parts initially. These approaches highlight the problems associated with trying to fabricate thick-section parts by injection moulding.

The critical needs for a new complex shape forming technology then are: separate the mould filling operation from the hardening process; make vehicle removal easier; and implement new setting mechanisms.

Gelcasting, methylcellulose gelation, agar gelation and reversible deflocculation overcome many of the problems associated with concurrent fluid flow and heat transfer. These processes inherently separate the mould filling operation from the setting operation. A fluid slurry is poured or injected into a mould using as much care as is required to make sure that all regions of the mould are filled properly. After filling, the slurry in the mould is gelled or flocculated to set the part in the mould. Therefore, defects such as jetting, weld

Fig. 6.19 Complicated parts can be made by the gelcasting process: (a) turbocharger rotor; (b) gear.

marks, voids, thermal strains, and flashing are avoided, typical parts made using the gelcasting process are shown in Fig 6.19.

All of the processes listed in Table 6.9 are solvent-based. After the part has been set in the mould and removed, the solvent can be removed in a drying step. Only a small amount of the dried part is binder, which is removed in a binder burnout step, similar to that used for dry pressed bodies. The use of a solvent-based system should be more amenable to making parts having thick and thin sections because vehicle removal is easier than in injection moulded parts.

Several new setting mechanisms are illustrated by the technologies listed in Table 6.9. The methylcellulose approach has been used successfully for metal powders [55]. The agarose gel technology has been used for alumina, but is limited to low solids loadings (< 40 vol% solids) [56]. The acrylamide gel system has been used for a wide range of materials and shapes [57, 58]. Pressure casting has shown significant promise in both the traditional ceramics industry and for prototype production of advanced ceramics [53, 63]. Freeze casting also has shown some promise for advanced ceramics [54]. Reversible deflocculation is an intriguing process [59], but has not been commercialized (M.M. Crutchfield, personal communication).

REFERENCES

1. Janney, M.A., Vance, M.C., Jordan, A.C. and Kertez, M.P. (1986) Bibliography of Ceramic Extrusion and Plasticity, *Report No. ORNL 6363*, Oak Ridge, TN.
2. Bagley, R.D. (1974) *Method for forming thinwalled honeycomb structures.* US Patent 3, 790, 654.
3. Onoda, Jr. G.Y. (1978) The rheology of organic binder solutions in *Ceramic Processing Before Firing* (eds G.Y. Onoda, Jr. and L.L. Hench) J. Wiley and Sons, NY.
4. Whittemore, J.W. (1944) Industrial use of plasticizers, binders, and other auxiliary agents. *Am. Ceram. Soc. Bull.*, **23**, 427–32.
5. McNamarra, E.P. and Comefora, J.E. (1945) Classification of natural organic binders. *J. Am. Ceram. Soc.*, **28**, 25–31.
6. Treischel, C.C. and Emrich, E.W. (1946) Study of several groups of organic binders under low pressure extrusion. *J. Am. Ceram. Soc.*, **29**, 129–32.
7. Wild, A. (1954) Review of organic binders for use in structural clay products. *Am. Ceram. Soc. Bull.*, **33**, 368–70.
8. Morse, T. (1979) *Handbook of Organic Additives for Use in Ceramic Body Formulation*, Montana MHD Research and Development Institute, Butte, MT.
9. Morawetz, H. (1965) *Macromolecules in Solution*, Wiley-Interscience, NY.
10. Yamakawa, H. (1971) *Modern Theory of Polymer Solutions*, Harper and Row, NY.
11. Elias, H.G. (1984) *Macromolecules*, Plenum Press, NY.
12. Schieffele, G.W. and Sacks, M.D. (1988) Pyrolysis of poly(vinylbutyral) binders, in *Ceramic Powder Science. II* (eds G.L. Messing, E.R. Fuller and H. Hausner) American Ceramic Society, Westerville, OH.
13. Anon, (1991) *Synthetics, Screening and Filtration Media*, Tetko, Inc., Briarcliff Manor, NY.
14. Anon. (1990) *Continuous Concept Screen Changers*, High Technology Corp., Hackensack, NJ.

15. Strivens, M.A. (1963) Injection molding of ceramic insulating materials. *Am. Ceram Soc. Bull.*, **42**, 13–19.
16. Janney, M.A. and Onoda, Jr. G.Y. (1987) Particulate mechanics of highly loaded ceramic systems. *Adv. Ceram.*, **21**, 615–26.
17. Goodson, F.J. (1959) Experiments in extrusion. *Trans. Br. Ceram Soc.*, **58**, 158–87.
18. Arakawa, M., Banerjee, S. and Williamson W.O.(1971) Extrusion behaviour of hard shale. *Am. Ceram. Soc. Bull.*, **50**, 933–5.
19. Janney, M.A. (1982) Plasticity of Ceramic Particulate Systems, PhD. Dissertation, University of Florida.
20. Atkinson, J.H. and Bransby P.L. (1978) *Mechanics of Soils*, McGraw-Hill, Maidenhead, UK.
21. Schofield, A.N. and Wroth C.P. (1968) *Critical State Soil Mechanics*, McGraw-Hill, Maidenhead, UK.
22. Russel, R. and Hanks, Jr. C.F (1942) Stress–strain characteristics of plastic clay masses. *J. Am. Ceram. Soc.*, **25**, 16–28.
23. Han, C.D. (1976) *Rheology in Polymer Processing*, Academic Press, NY.
24. Mosthaf, E. (1949) Long extrusion dies cut losses. *Ceram. Ind* (Sevres, France) **53**, 70.
25. Butterworth, B., Baldwin, L.W. and Coley, S.G. (1966) Dies for extruding perforated bricks. *J. Br. Ceram. Soc.*, **3**, 563.
26. Lachman, I.M., Bagley, R.D. and Lewis R.M. (1981) Thermal expansion of extruded cordierite ceramics. *Am. Ceram. Soc. Bull.*, **60**, 202.
27. Robinson, G. (1978) Extrusion defects, in *Ceramic Processing Before Firing* (eds G.Y. Onoda, Jr. and L.L. Hench) J. Wiley and Sons, NY.
28. Robinson, G.C., Kizer, R.H. and Duncan, J.F. (1968) Raw material parameters determining extrudability. *Am. Ceram. Soc. Bull.*, **47**, 822–32.
29. Seanor, J.G. and Schweltzer, W.P. (1962) Basic theoretical factors in extrusion augers. *Am. Ceram. Soc. Bull.*, **41**, 560–63.
30. Lund, H.H., Bortz, S.A. and Reed, A.J. (1962) Auger design for clay extrusion. *Am. Ceram. Soc. Bull.*, **41**, 554–9.
31. Parks, J.R. and Hill, M.J. (1950) Design of extrusion augers and the characteristic equation of ceramic extrusion machines. *J. Am. Ceram. Soc.*, **42**, 1–6.
32. Tarpley, W.B., Yocum, K.H. and Pheasant, R. (1961) Ultrasonic Extrusion: Reduction in Vehicle and Plasticizer Requirements for Non-Clay Ceramics, *USAEC Report No. NYO–10006*, New York.
33. Zavanovic, B.M. and Janjic, O.G. (1978) Cold forming of heavy clay products by the double layer technique. *Ceramurgia*, **8**, 201– 6.
34. Frados, J. (ed) (1976) *Plastics Engineering Handbook*, Van Nostrand Reinhold, NY.
35. Rubin, I. (1973) *Injection Molding of Plastics*, J. Wiley and Sons, NY.
36. Haglund, T.R. (1936) *Process of producing refractory*. US Patent 2, 048, 861.
37. Schwartzwalder, K. (1949) Injection molding of ceramic materials. *Am. Ceram. Soc. Bull.*, **28**, 459–61.
38. Edirisinghe, M.J. and Evans, J.R.G. (1986) Review: fabrication of engineering ceramics by injection molding. I. Materials selection. *Int. J. High Technol. Ceram.*, **2**, 1–31.
39. Edirisinghe, M.J. and Evans, J.R.G. (1986) Review: fabrication of engineering ceramics by injection molding. II. Techniques. *Int. J. High Technol. Ceram.*, **2**, 249–78.
40. White, J.L. and Dee, H.B. (1974) Flow visualization for injection molding of polyethylene and polystyrene melts. *Polym. Eng. Sci.*, **14**, 212–22.
41. Edirisinghe, M.J. (1991) Fabrication of engineering ceramics by injection molding. *Am. Ceram. Soc Bull.*, **70**, 824.

42. Bandyopadhyay, G. and French, K.W. (1994) Effect of powder characteristics on injection molding and burnout cracking. *Am. Ceram. Soc. Bull.*, **73**, 107–14
43. Horton, R.A. (1974) *Molded refractory articles*, US Patent 3, 859,405.
44. Prokaev, V.P., Kashirnikov V.M. (1976) *Ceramic mixture, containing ethyl silicate, gel inducer, zircon, and disthene-silimanite, for producing injection moulds and rods, primarily for casting aluminum alloys.* USSR Patent SU 535131.
45. Pratilio, F. (1985) *Equipment Catalogue*, Educational Machinery Corporation, Greenwich, CT.
46. Pasto, A.E. and Natansohn, S. (1991) Development of Improved processing for High Reliability Structural Ceramics for Advanced heat Engines. Final Report, *GTE Labs TR-0172-12-91-800*, Oak Ridge, TN, USA.
47. Zhang, H., German, R.M. and Bose, A. (1990) Wick debinding distortion of injection molded powder compacts. *Int. J. Powder Met.*, **26**, 217.
48. Wright, J.K. and Evans, J.R.G. (1991) Removal of organic vehicle from moulded ceramic bodies by capillary action. *Ceram. Int.*, **17**, 9–87.
49. Mutsuddy, B.C. (1983) Injection molding research paves way to ceramic engine parts. *Ind. Res. Dev.*, July 76–7.
50. Peltsman, M. (1986) Low pressure injection moulding and mould design. *MPR*, May 367–9.
51. German, R.M. and Hens, K.F. (1991) Key issues in powder injection molding. *Am. Ceram. Soc. Bull*, **70**, 1294.
52. Zhang, J.G., Edirisinghe, M.J. and Evans, J.R.G. (1989) A catalogue of ceramic injection moulding defects and their causes. *Ind. Ceram.*, **9**, 72.
53. Blanchard, E.G. (1991) Pressure casting improves productivity. *Am. Ceram. Soc. Bull.*, **67**, 1680–83.
54. Novich, B.E., Sundback, C.A. and Adams, R.W (1992) Quickset injection molding of high performance ceramics. *Ceram. Trans.*, **26**, 157–64.
55. Rivers, R.D. (1976) *Method of injection molding powder metal parts*, US Patent 4, 113,480.
56. Fanelli, A.J, Silvers, R.D., Frei, W.S., Burlew, J.V. and Marsh, G.B. (1989) New aqueous injection molding process for ceramic powders. *J. Am. Ceram. Soc.*, **72**, 1833–3.
57. Young, A.C., Omatete, O.O., Janney, M.A. and Menchhofer, P.A. (1991) Gelcasting of alumina *J. Am. Ceram. Soc.*, **74**, 612–16.
58. Omatete, O.O., Janney, M.A. and Strehlow, R.A. (1991) Gelcasting – a new ceramic forming process. *Am. Ceram. Soc. Bull.*, **70**, 1641–6.
59. Crutchfield, M.M. (1982) *Reversible deflocculation of clay slurries*, US. Patent 4, 327,189.
60. Johnson, A., Carlstrom, E., Heimansson, L. and Carlsson, R. (1984) Rate-controlled thermal extraction of organic binders from injection-molded bodies, in *Advances in Ceramics, Vol. 9, Forming of Ceramics* (eds J.A. Mangels and G.L. Messing) American Ceramic Society, Columbus, OH.
61. Bandyopadhyay, G., French, K.W., Bowen, L.J. and Neil, J.T. (1986) *Large cross section injection molded ceramic shapes*, Eur. Pat. Appl. EP 196600.
62. Miyauchi, J. and Kobayashi, Y. (1985) Development of silicon nitride turbine rotors *SAE Tech. Pap. Ser.*, No 850313.
63. Pujari, V.K., Amin, K.E. and Tewari, P.H. (1991) Development of improved processing and evaluation of silicon nitride, in *Proc. 28th ATD-CCM*, SAE P243, April, 1991. Society of Automotive Engineers, Detroit, MI, USA.

Index

Page numbers appearing in **bold** refer to figures and page numbers appearing in *italic* refer to tables.